Mold for cookies

Crowded station

Measure per unit quantity

Maple-leaf-shaped cake

Angles of Figures

Patterns by tessellation

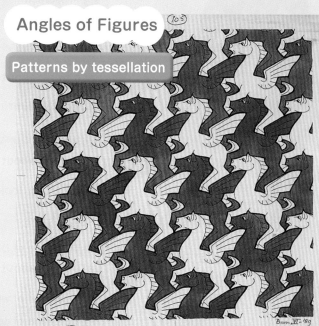

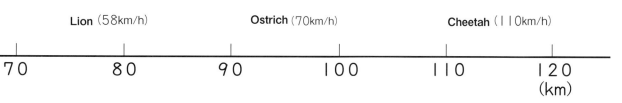

Lion (58km/h)　　　Ostrich (70km/h)　　　Cheetah (110km/h)

70　　　80　　　90　　　100　　　110　　　120

(km)

Table of contents

Hiroto

Nanami

Yui

Let's learn mathematics together.

Daiki

5th Grade. Volume 2

3 learning competencies to develop

1 Thinking Competency

Competency to think the same or similar way

Competency to find what you have learned before and think in the same or similar way

Competency to find rules

Competency to analyze numbers and various expressions that change together, and investigate any rules

Competency to explain the reason

Competency to explain the reason why, based on learned rules and important ideas

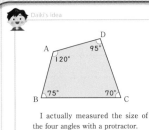

Daiki's idea

I actually measured the size of the four angles with a protractor.

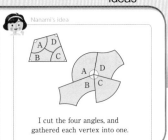

Nanami's idea

I cut the four angles, and gathered each vertex into one.

2 Judgement Competency

Competency to find mistakes

Competency to find the over generalization of conventions and rules

Competency to categorize by properties and patterns

Competency to categorize numbers and shapes by observing their properties and patterns

Competency to compare ideas and ways of thinking

Competency to find the same or different ways of thinking by comparing friends' ideas and your own ideas

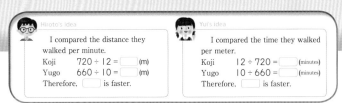

Hiroto's idea

I compared the distance they walked per minute.
Koji $720 \div 12 = \boxed{}$ (m)
Yugo $660 \div 10 = \boxed{}$ (m)
Therefore, $\boxed{}$ is faster.

Yui's idea

I compared the time they walked per meter.
Koji $12 \div 720 = \boxed{}$ (minutes)
Yugo $10 \div 660 = \boxed{}$ (minutes)
Therefore, $\boxed{}$ is faster.

3 Representation Competency

Competency to represent sentences with diagrams or expressions

Competency to read problem sentences, draw diagrams, and represent them using expressions

Competency to represent data in graphs or tables

Competency to express the explored data in tables or graphs, in simpler and easier ways, according to the purpose

Competency to communicate with friends and yourself

Competency to communicate your ideas to friends in simpler and easier manners and to write notes in simpler and easier ways for you

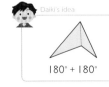

Daiki's idea

$180° + 180°$

Nanami's idea

$180° + 360° - 180°$

Let's find monsters.

Monsters
which represent ways of thinking in mathematics

Setting the unit.

Once you have decided one unit, you can represent how many.

Unit

If you try to arrange...

You can compare if you align the number place and align the unit.

Align

If you try to separate...

Decomposing numbers by place value and dividing figures make it easier to think about problems.

Separate

If you try to summarize...

It is easier to understand if you put the numbers in groups of 10 or summarize in a table or graph.

Summarize

If you represent in other way...

If you represent in other expression, diagram, table, etc., it is easier to understand.

Other way

If you try to change the number or figure...

If you try to change the problem a little, you can better understand the problem or find a new problem.

Change

Can you do the same or similar way?

If you find something the same or similar, then you can understand.

Looks same

Is there a rule?

If you examine a few examples, then you can find out whether there is a rule.

Rule

You wonder why?

Why this happens? If you communicate the reasons in order, it will be easier to understand for others.

Why

Ways to think learned in the 4th grade.

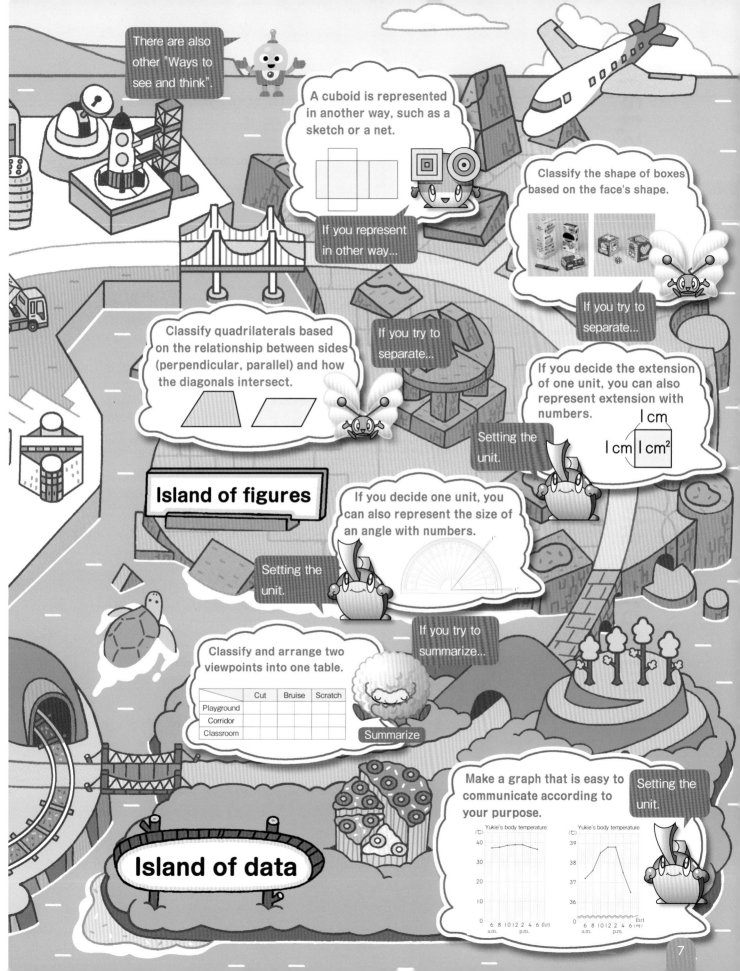

There are also other "Ways to see and think".

A cuboid is represented in another way, such as a sketch or a net.

If you represent in other way...

Classify the shape of boxes based on the face's shape.

If you try to separate...

Classify quadrilaterals based on the relationship between sides (perpendicular, parallel) and how the diagonals intersect.

If you try to separate...

If you decide the extension of one unit, you can also represent extension with numbers.

1 cm
1 cm 1 cm²

Setting the unit.

Island of figures

If you decide one unit, you can also represent the size of an angle with numbers.

Setting the unit.

If you try to summarize...

Classify and arrange two viewpoints into one table.

	Cut	Bruise	Scratch
Playground			
Corridor			
Classroom			

Summarize

Make a graph that is easy to communicate according to your purpose.

Setting the unit.

Yukie's body temperature

Yukie's body temperature

Island of data

Let's progress through 3 ways of learning.

 Self directed learning: Learning on your own initiative.

Find the ? Problem

If you find a problem in your life or mathematics, you will like to solve it.

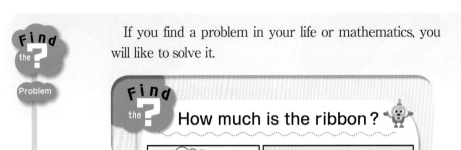

Find the ? How much is the ribbon?

Want to solve
Want to think
Want to know
Want to explore

You will like to represent the problem with mathematical diagrams and expressions. The "purpose" is born.

Want to think Multiplication in vertical form

2 Let's think about how to calculate 80 x 2.4 in vertical form.

```
    8 0
×   2.4
```

Can I use the method for decimal number × whole number?   Is the position of the decimal point the same as in decimal number × whole number?

Diaki   Nanami

Want to compare
Want to represent
Want to discuss
Want to improve
Want to communicate

Compared to your own ideas, search for good or similar ideas from your friends.

Want to explain

③ Let's explain the calculation methods of the following children.

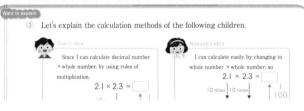

Diaki's idea
Since I can calculate decimal number × whole number, by using rules of multiplication.

$2.1 \times 2.3 =$ ☐

Nanami's idea
I can calculate easily by changing to whole number × whole number, so

$2.1 \times 2.3 =$ ☐

10 times 10 times | 100

Want to confirm

Make sure to confirm that the "summary" can be used for other problems.

Want to confirm

2 In the following calculations, let's place the decimal point of the product.

```
①    5.6        ②   3.27       ③    1.48
    × 4.3           ×  1.2           ×  2.5
    1 6 8           6 5 4            7 4 0
    2 2 4           3 2 7            2 9 6
```

Want to try

When you can do it, you will want to have more problems.

Want to try

3 Let's solve the following calculations in vertical form.

① 0.2 × 1.6   ② 0.4 × 0.35

```
      0.2                0.4
×   1.6              × 0.3 5
```

⭐2 Dialogue learning: Learning together with friends.

As learning progresses, you will want to know what your friends are thinking. Also, you will like to share your own ideas to your friends. Let's discuss next to each other, in groups, or with the whole class.

Want to discuss

I draw a straight line with the same length as side BC.

After, all you need to know is the position of vertex A.

Want to explain

① Let's find the appropriate number for ☐ .
② Let's create a multiplication problem by changing the numbers and words.
③ Let's create a division problem by changing the numbers and words.

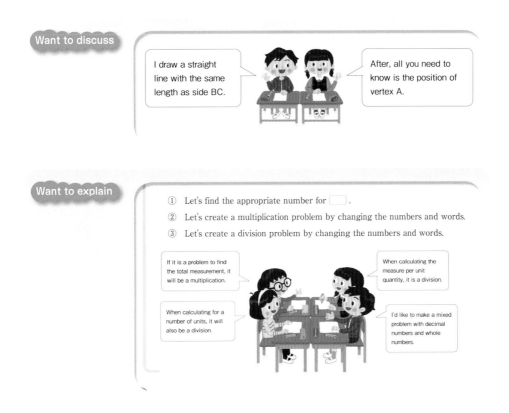

If it is a problem to find the total measurement, it will be a multiplication.

When calculating the measure per unit quantity, it is a division.

When calculating for a number of units, it will also be a division.

I'd like to make a mixed problem with decimal numbers and whole numbers.

⭐3 Deep learning: Usefulness and efficiency of what you learned.

Let's cherish the feeling "I want to know more." and "Can I do this in another case?" Let's deepen learning in life and mathematics.

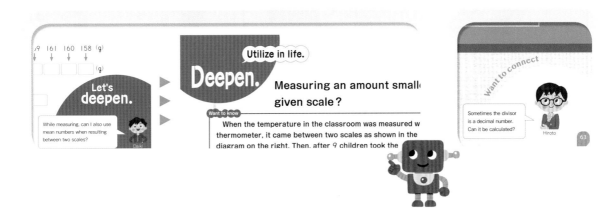

Let's deepen.

While measuring, can I also use mean numbers when resulting between two scales?

Utilize in life.

Deepen. Measuring an amount small given scale?

Want to know

When the temperature in the classroom was measured w thermometer, it came between two scales as shown in the diagram on the right. Then, after 9 children took the

Sometimes the divisor is a decimal number. Can it be calculated?

Hiroto

63

What are the differences between a whole number and a decimal number?

Elevation of Malaita Island   1 4 3 5 [ ]

Rail width   1 4 3 5 [ ]

Both became 1 435, but the unit is different.

The elevation of Malaita Island is 1435 m.

The rail width is 1435 mm.

If the units are aligned, it will become 1 435 m and 1.435 m.

The way the numerals are arranged is the same, but the size is completely different.

1

2

3

Problem   Is there any relationship between the structure of whole numbers and decimal numbers?

Decimal Numbers and Whole Numbers

Let's explore the structure and the size of numbers.

Let's explore about the numbers 1435 and 1.435.

 Hiroto

The arrangement of numbers is the same.

The size of each number is different.

Yui

Purpose Is the structure of whole numbers the same as of decimal numbers?

Want to represent

① Let's write each number in the following table.

	Thousands	Hundreds	Tens	Ones	$\frac{1}{10}$	$\frac{1}{100}$	$\frac{1}{1000}$	
Elevation of Malaita Island								m
Rail width								m

② Let's fill in the ☐ with numbers.

1435 gathers 1 set of ☐ , 4 sets of ☐ , 3 sets of ☐ ,

and 5 sets of ☐ .

1.435 gathers 1 set of ☐ , 4 sets of ☐ , 3 sets of ☐ ,

and 5 sets of ☐ .

③ Let's represent each number using math sentences.

1435 = 1000 + 400 + 30 + 5

= 1000 × ☐ + 100 × ☐ + 10 × ☐ + 1 × ☐

1.435 = 1 + 0.4 + 0.03 + 0.005

= 1 × ☐ + 0.1 × ☐ + 0.01 × ☐ + 0.001 × ☐

Way to see and think

Think about how many sets you have for each place value.

1 Let's explore the structure of numbers.

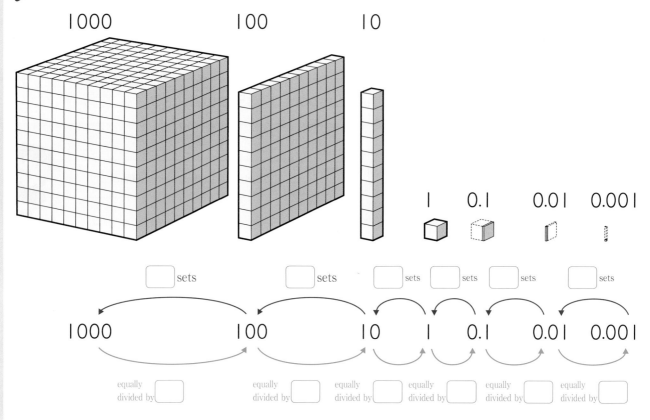

1000   100   10

1   0.1   0.01   0.001

☐ sets   ☐ sets   ☐ sets   ☐ sets   ☐ sets   ☐ sets

1000   100   10   1   0.1   0.01   0.001

equally divided by ☐   equally divided by ☐   equally divided by ☐   equally divided by ☐   equally divided by ☐   equally divided by ☐

💡 Summary

For both whole numbers and decimal numbers, if 10 sets are gathered then the number is moved to the next higher place value, and if the number is equally divided by 10 (equivalently $\frac{1}{10}$) then the number is moved to the next lower place value.

Following this, any whole number or decimal number can be represented by using the ten numerals 0, 1, 2, …, 9 and the decimal point.

 2 Let's create the following numbers by using the ten numerals from 0 to 9 (only once), and the decimal point.

① The smallest number.

② The number that is less than 1 but closest to 1.

2 Let's investigate about 10 times, 100 times, and 1000 times of 1.342.

Purpose What number becomes when you calculate 10 times, 100 times, and 1000 times of a number?

① Let's find 10 times, 100 times, and 1000 times of 1.342, and write it down in the following table.

Thousands	Hundreds	Tens	Ones	$\frac{1}{10}$	$\frac{1}{100}$	$\frac{1}{1000}$
			1	3	4	2
10 times of 1.342 →						
100 times of 1.342 →						
1000 times of 1.342 →						

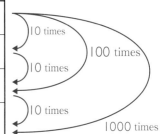

② If you calculate 10 times, 100 times, and 1000 times of a number, how does the place value of the number change?

③ Let's fill in the ☐ on the right with the decimal point when you calculate 10 times, 100 times, and 1000 times of a number.

1 . 3 4 2
1 ☐ 3 ☐ 4 ☐ 2 ☐
1 ☐ 3 ☐ 4 ☐ 2 ☐
1 ☐ 3 ☐ 4 ☐ 2 ☐

Way to see and think

If you align and write the numerals, it is easy to understand the differences on the decimal point position.

Summary

When you identify 10 times, 100 times, 1000 times, ..., of a number, the number's decimal point moves respectively 1 place, 2 places, 3 places, ..., to the right.

④ Let's try to represent with math sentences 10 times, 100 times, and 1000 times of 1.342.

ⓐ 1.342 × 10 = ☐   ⓑ 1.342 × 100 = ☐

ⓒ 1.342 × 1000 = ☐

 Let's answer the following questions.

① Let's write 10 times, 100 times, and 1000 times of 23.47.

② How many times of 8.72 do 87.2, 872, and 8720 represent respectively?

3 **Let's explore about $\frac{1}{10}$ and $\frac{1}{100}$ of 125.**

⊙ Purpose What number becomes when you calculate $\frac{1}{10}$ and $\frac{1}{100}$ of a number?

① Let's find $\frac{1}{10}$ and $\frac{1}{100}$ of 125, and write it down in the following table.

Hundreds	Tens	Ones	$\frac{1}{10}$	$\frac{1}{100}$
1	2	5		

$\frac{1}{10}$ of 125 →

$\frac{1}{100}$ of 125 →

As for the number with size $\frac{1}{10}$ of 125, since
$100 \div 10 = 10$,
$20 \div 10 = 2$,
$5 \div 10 = 0.5$,
and $10 + 2 + 0.5 = 12.5$,
then the answer is 12.5.

② If you calculate $\frac{1}{10}$ and $\frac{1}{100}$ of a number, how does the place value of the number change?

③ Let's fill in the ☐ on the right with the decimal point when you calculate $\frac{1}{10}$ and $\frac{1}{100}$ of 125.

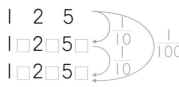

1 2 5
1☐2☐5☐
1☐2☐5☐

Way to see and think

If you align and write the numerals, it is easy to understand the differences on the decimal point position.

⊙ Summary

When you identify $\frac{1}{10}$, $\frac{1}{100}$, …, of a number, the number's decimal point moves respectively 1 place, 2 places, …, to the left.

④ Let's try to represent with math sentences $\frac{1}{10}$ and $\frac{1}{100}$ of a number.

ⓐ $125 \div 10 = \boxed{}$ ⓑ $125 \div 100 = \boxed{}$

⑤ What number becomes $\frac{1}{1000}$ of 125?

4 ▶ Let's answer the following questions.

① Let's write $\frac{1}{10}$ and $\frac{1}{100}$ of 30.84.

② What fraction of 63.2 do 6.32 and 0.632 represent respectively?

Notebook for thinking

Let's write how you thought.

Write today's date.

April 18

Let's write 10 times, 100 times, and 1000 times of 1.342.

Write the problem of the day that you must know.

Ten Thousands	Thousands	Hundreds	Tens	Ones	$\frac{1}{10}$	$\frac{1}{100}$	$\frac{1}{1000}$
				1 .	3	4	2
			1	3 .	4	2	
		1	3	4 .	2		
		1	3	4	2		

10 times of 1.342

100 times of 1.342

1000 times of 1.342

10 times
10 times
10 times
100 times
1000 times

Summary

If you calculate 10 times, 100 times, and 1000 times of a number, the number's decimal point moves respectively to the right.

To make it easy to understand, the summary can be enclosed in a line and using colors.

⟨Reflect⟩

○ In a table, if you calculate 10 times, 100 times, and 1000 times of a number, then you can see as the numerals move to the left respectively.

○ If you look at the decimal point, you can see as it moves to the right respectively.

As for reflection, write the following:
● things you understood,
● interesting thoughts,
● doubts,
● and what you want to inquire more.

What you can do now

☐ **Understanding that whole numbers and decimal numbers have the same structure.**

1 Let's summarize what is common to whole numbers and decimal numbers.

① Both whole numbers and decimal numbers are represented by the same place system: when [ ] sets are gathered, the place value increases by one, and by equally dividing in [ ] parts the place value decreases by one.

② Any whole number or decimal number can be represented by using the [ ] numerals from 0 to 9 and a decimal point.

☐ **Can represent the structure of whole numbers and decimal numbers with math sentences.**

2 Let's fill in the [ ] with numbers.

① $8617 = \boxed{} \times 8 + \boxed{} \times 6 + \boxed{} \times 1 + \boxed{} \times 7$

② $86.17 = \boxed{} \times 8 + \boxed{} \times 6 + \boxed{} \times 1 + \boxed{} \times 7$

③ $0.8617 = \boxed{} \times 8 + \boxed{} \times 6 + \boxed{} \times 1 + \boxed{} \times 7$

☐ **Understanding 10 times, 100 times, 1000 times, $\frac{1}{10}$, and $\frac{1}{100}$ of a number.**

3 Let's find the following numbers.

① 10 times, 100 times, and 1000 times of 5.67

② $\frac{1}{10}$ and $\frac{1}{100}$ of 596

③ 0.95×10 ④ 0.95×100 ⑤ 0.95×1000

⑥ $36.7 \div 10$ ⑦ $36.7 \div 100$

☐ **Can create numbers by using the structure of numbers.**

4 Let's write the number closest to 30 by using one time the five numerals 2, 3, 4, 8, 9 and a decimal point.

Supplementary problems
p.146

Usefulness and efficiency of learning

1 Let's fill in the ☐ with numbers.

① $3805 = 1000 \times \boxed{} + 100 \times \boxed{} + 10 \times \boxed{} + 1 \times \boxed{}$

② $38.05 = 10 \times \boxed{} + 1 \times \boxed{} + 0.1 \times \boxed{} + 0.01 \times \boxed{}$

③ $0.3805 = 0.1 \times \boxed{} + 0.01 \times \boxed{}$
$+ 0.001 \times \boxed{} + 0.0001 \times \boxed{}$

☐ Understanding that whole numbers and decimal numbers have the same structure.

2 Let's answer the following questions.

① How many times of 0.472 is represented in each number below?

ⓐ 47.2　　　ⓑ 472　　　ⓒ 4.72

② What fraction of 61.6 is represented in each number below?

ⓐ 0.616　　　ⓑ 6.16

③ Let's write 10 times, 100 times, and 1000 times of 5.93.

④ Let's write 10 times, 100 times, and 1000 times of 0.082.

⑤ Let's write $\frac{1}{10}$ and $\frac{1}{100}$ of 0.75.

⑥ Let's write $\frac{1}{10}$ and $\frac{1}{100}$ of 30.1.

☐ Can represent the structure of whole numbers and decimal numbers with math sentences.

☐ Understanding 10 times, 100 times, 1000 times, $\frac{1}{10}$, and $\frac{1}{100}$ of a number.

3 There are six cards with the numerals 0, 1, 5, 6, 9 and a decimal point. Let's not place the 0 and the decimal point at the end, and use all the cards to create the following numbers.

$\boxed{0}\ \boxed{1}\ \boxed{5}$
$\boxed{6}\ \boxed{9}\ \boxed{.}$

① The smallest number.

② The largest number.

③ The number that is larger than 1 but closest to 1.

④ The number that is smaller than 6 but closest to 6.

☐ Can create numbers by using the structure of numbers.

Which one is the same?

Problem

How can you explore figures that overlap exactly?

18

2 Congruent Figures

Let's explore the properties of figures with the same shape and size and how to draw them.

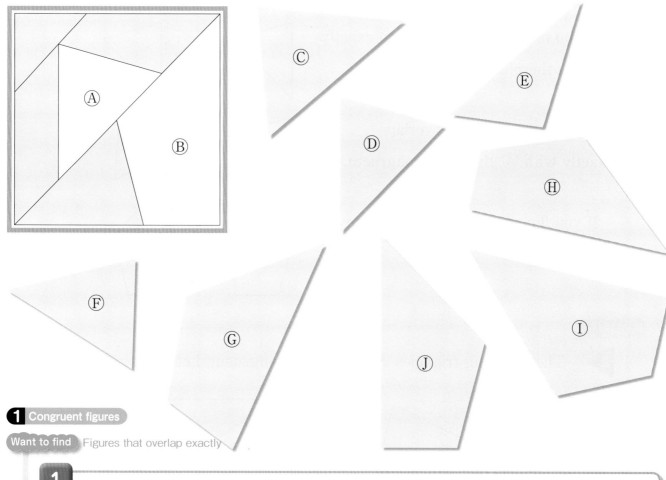

 1 Congruent figures

Want to find Figures that overlap exactly

1 Let's find a figure from Ⓒ～Ⓕ that exactly overlaps with triangle Ⓐ.

Hiroto

If you look at it, you don't immediately know which one.

Which part should I look at to find that the size and shape of two figures are the same?

Nanami

Want to try

Let's cut out the figures on p.162 and try to explore.

1 Let's find a figure from Ⓖ～Ⓙ that exactly overlaps with quadrilateral Ⓑ.

Let's explore by moving, turning, and flipping over the figure.

Two figures are said to be **congruent** when both overlap exactly.

Congruent figures have the same shape and size.

If you move Ⓓ, it will overlap exactly with Ⓐ, therefore congruent.

If you turn Ⓕ, it will overlap exactly with Ⓐ, therefore congruent.

If you flip Ⓙ, it will overlap exactly with Ⓑ, therefore congruent.

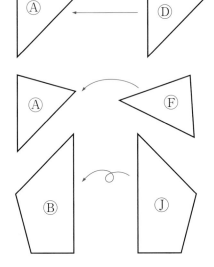

Want to find

2 The following triangles Ⓚ and Ⓛ are congruent. Let's answer the questions below.

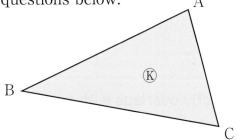

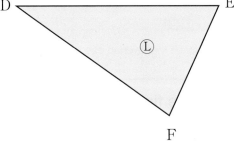

① Let's say all the vertices that overlap.

② Let's say all the sides that overlap.

③ Let's say all the angles that overlap.

In congruent figures, overlapping vertices, overlapping sides, and overlapping angles are called **corresponding vertices, corresponding sides**, and **corresponding angles** respectively.

3 The quadrilaterals Ⓜ and Ⓝ, shown on the right, are congruent. Let's say all the corresponding vertices, sides, and angles.

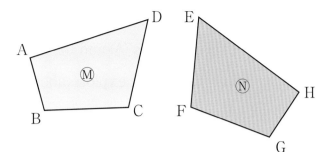

Properties of congruent figures

 2 Let's compare the length of the corresponding sides and the size of the corresponding angles from triangles Ⓚ and Ⓛ on exercise ②. What do you understand?

What does it mean to overlap exactly?

Daiki

Can I identify if I measure the length of the sides and the size of the angles?

Yui

🌱**Purpose** What are the properties of congruent figures?

① Let's measure and compare the length of corresponding sides.

② Let's measure and compare the size of corresponding angles.

💡**Summary**

In congruent figures, the corresponding sides have equal length. Also, the corresponding angles have equal size.

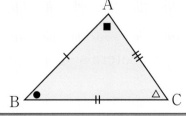 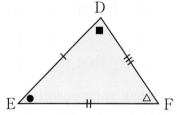

 4 Let's compare the length of the corresponding sides and size of the corresponding angles in quadrilaterals Ⓜ and Ⓝ from exercise ③.

5 The triangles Ⓐ and Ⓑ, shown on the right, are congruent. Let's explore about corresponding sides and angles.

① Which is the corresponding side to side DF? Also, how many cm is the length of side DF?

② Which is the corresponding angle to angle E? Also, how many degrees is the size of angle E?

③ Let's find the length of the other sides and size of the other angles in triangle Ⓑ.

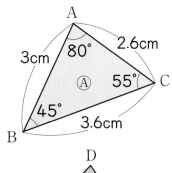

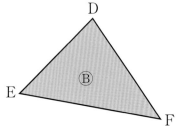

6 The quadrilaterals Ⓒ and Ⓓ, shown on the right, are congruent. Let's explore about corresponding sides and angles.

① Which is the corresponding side to side GF? Also, how many cm is the length of side GF?

② Which is the corresponding angle to angle E? Also, how many degrees is the size of angle E?

③ Let's find the length of the other sides and size of the other angles in quadrilateral Ⓓ.

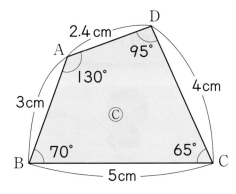

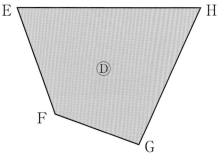

7 Can you say that the following quadrilaterals Ⓔ and Ⓕ are congruent? Let's also say the reasons.

Way to see and think

Which properties of congruent figures are present?

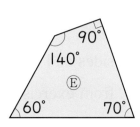

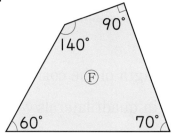

Activity

1

Let's think about how to draw a triangle that is congruent to the triangle shown on the right.

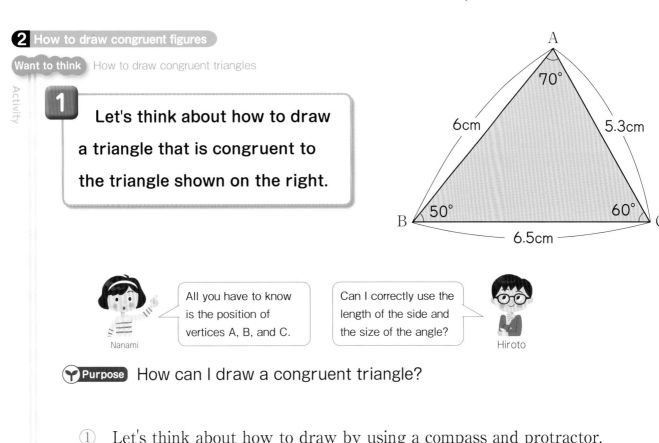

All you have to know is the position of vertices A, B, and C.

Nanami

Can I correctly use the length of the side and the size of the angle?

Hiroto

🎯 **Purpose** How can I draw a congruent triangle?

① Let's think about how to draw by using a compass and protractor.

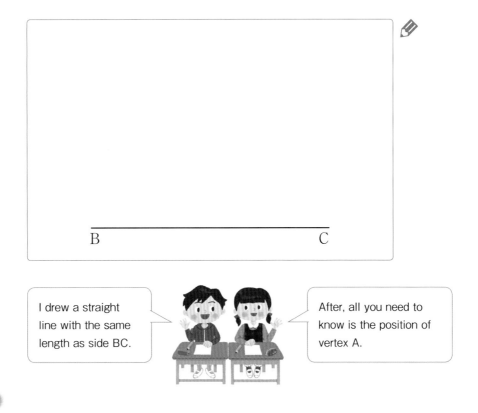

I drew a straight line with the same length as side BC.

After, all you need to know is the position of vertex A.

Want to discuss

② Let's discuss how to determine the position of vertex A.

Measure the length of the 3 sides, and then draw them.

B   C

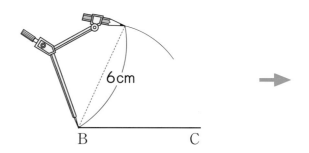

Measure the length of 2 sides and the angle in between, and then draw them.

B   C

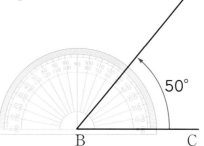

Measure the length of 1 side and the angles formed by that side with the other two sides, and then draw them.

B   C

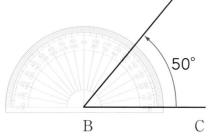

Want to explain

③ Let's explain the procedure followed by the 3 children on each drawing method.

Want to confirm

④ Let's confirm that the drawn triangle ABC is congruent to the triangle from **1** in the previous page.

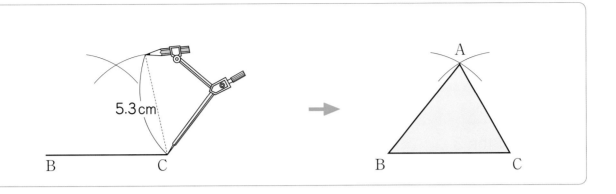

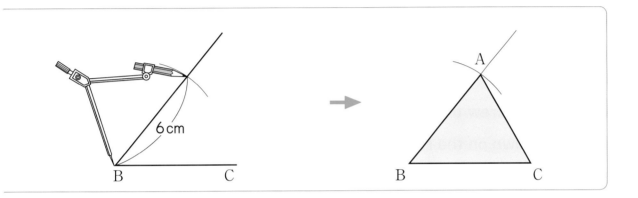

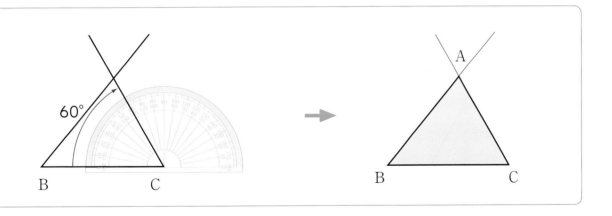

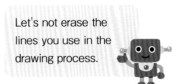

Let's not erase the lines you use in the drawing process.

🔵 Summary

Among the length of the 3 sides and the size of the 3 angles, if one of the following Ⓐ, Ⓑ, and Ⓒ is known, then a congruent triangle can be drawn.

Ⓐ Length of the 3 sides.

Ⓑ Length of 2 sides and the angle in between.

Ⓒ Length of 1 side and the angles formed by that side with the other two sides.

Want to discuss

2

Daiki and Yui drew triangle ABC under the conditions shown on the right. Let's discuss the reasons why the drawings from the 2 children are different.

- Length of side AB is 4 cm
- Length of side BC is 4.4 cm
- Size of angle C is 60°

Way to see and think

Let's try to think the reasons why this happens.

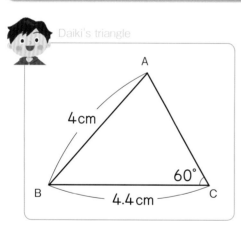

Daiki's triangle

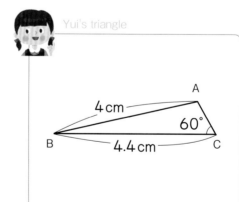

Yui's triangle

When you know the size of the 3 angles, can you draw a congruent triangle?

Want to confirm

 Let's measure the required length of the sides or size of the angles and draw a congruent triangle to the triangle shown on the right. Also, let's explain how you drew it.

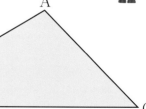

26

Activity

3 Let's think about how to draw a quadrilateral that is congruent to the following quadrilateral.

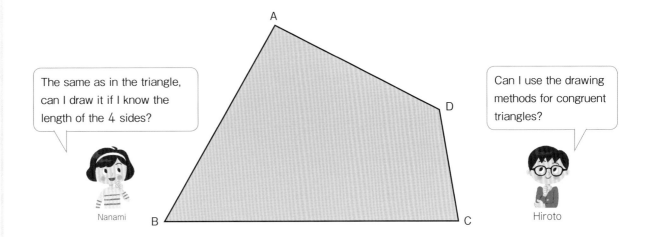

The same as in the triangle, can I draw it if I know the length of the 4 sides?

Nanami

Can I use the drawing methods for congruent triangles?

Hiroto

Purpose How can I draw a congruent quadrilateral?

Want to find

① Can you draw a congruent quadrilateral only using the length of the 4 sides?

 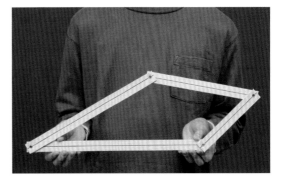

② Let's measure the length of the sides and the size of the angles from the above figures, and draw congruent quadrilaterals in your notebook.

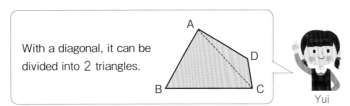

With a diagonal, it can be divided into 2 triangles.

Yui

Thinking based on the drawing methods for congruent triangles.

③ Let's explain the drawing methods of the following children.

 Hiroto's drawing method

Determine vertex D using the idea from triangles.
Measure the length of side AD and side CD.

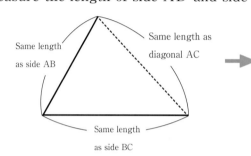

Same length as side AB

Same length as diagonal AC

Same length as side BC

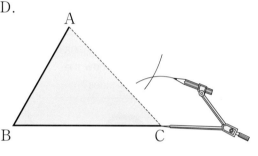

 Nanami's drawing method

Determine vertex D using the idea from triangles.
Measure the angles formed by diagonal AC and the other two sides.

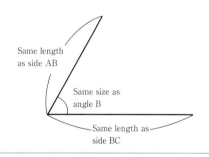

Same length as side AB

Same size as angle B

Same length as side BC

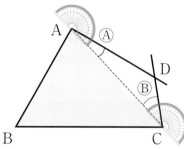

Summary

Congruent quadrilaterals can be drawn by using the drawing methods from congruent triangles if the quadrilateral is divided into two triangles by a diagonal.

 2 Let's draw a quadrilateral that is congruent to the quadrilateral shown on the right.

 Which sides and angles shall we use?

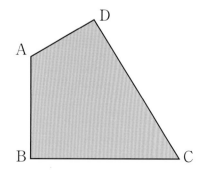

4 Let's tessellate congruent triangles like the diagram shown below. Also, let's discuss what you notice.

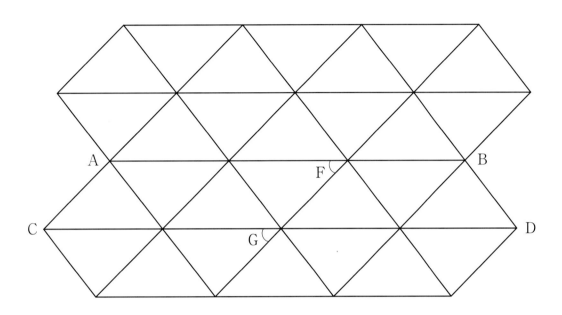

① Is the size of angle F and angle G equal?

② Let's confirm, using triangle rulers, that straight line AB and straight line CD are parallel.

③ Let's find parallelograms in the above diagram, and explain how each become a parallelogram.

④ Let's find trapezoids in the above diagram and explain how each become a trapezoid by using the words "parallel" and "straight line."

Way to see and think

Let's clarify the reasons and explain in order.

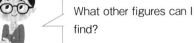

What other figures can I find?

Hiroto

What you can do now

☐ **Understanding corresponding vertices, sides, and angles of congruent figures.**

1 The triangles Ⓐ and Ⓑ are congruent. Let's answer the following questions.

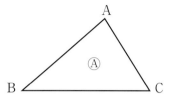
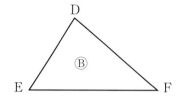

① Let's say all the corresponding vertices.

② Let's say all the corresponding sides.

③ Let's say all the corresponding angles.

☐ **Can draw congruent triangles.**

2 Let's draw a triangle that is congruent to the following triangles.

① Triangle that has sides with length 4 cm, 7 cm, and 8 cm.

② Triangle that has two sides with length 5 cm and 8 cm, and the angle in between with size 75°.

③ Triangle with two angles of size 45° and 60°, and the side in between with length 6 cm.

④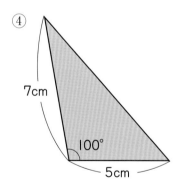

⑤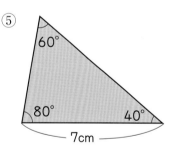

☐ **Can draw congruent quadrilaterals.**

3 Let's draw a rhombus that is congruent to the rhombus on the right.

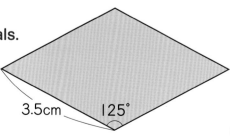

Supplementary problems ••••••••➤ p.147

Usefulness and efficiency of learning

1 The following two quadrilaterals are congruent. Let's say the length of all the corresponding sides and the size of all the corresponding angles.

Understanding corresponding vertices, sides, and angles of congruent figures.

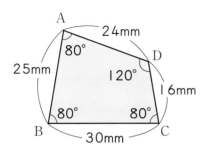
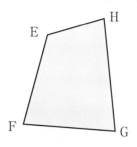

2 Let's measure the length of the corresponding sides and size of the corresponding angles, and draw figures that are congruent to the following figures in your notebook.

Can draw congruent triangles.

① Equilateral triangle

② Isosceles triangle

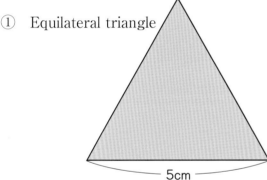
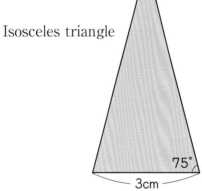

3 Draw a quadrilateral that is congruent to the quadrilateral on the right.

Can draw congruent quadrilaterals.

Let's explain the ideas of the following children, and draw with each of the drawing methods.

【Nanami's idea】

As shown in the figure on the right, if the quadrilateral is divided in 2, the triangle ABC is an isosceles triangle. Therefore, if you measure the size of angle Ⓐ, then you can draw a congruent quadrilateral.

【Hiroto's idea】

Even if it is not divided into 2 triangles, if you use right angles, then you can draw a congruent quadrilateral.

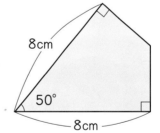
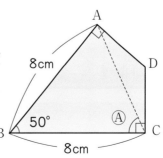

What are the rules for two quantities changing together?

Let's create a rectangular flowerbed with a surrounding length of 30 m.

If you change the length, then the area also changes.

If you increase the length, then the width decreases.

1

Let's enclose the surrounding with bricks.

The number of bricks increases as you pile up the first and second level.

If you add more bricks, the height of the enclosure will increase.

2

Problem There are various quantities that change together, but is there any rule?

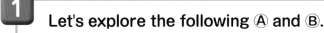

3 Proportion
Let's explore quantities changing together and its correspondence.

1 Quantities changing together

Want to explore

1

Let's explore the following Ⓐ and Ⓑ.

Ⓐ The length and width of a rectangle with a surrounding length of 30 m.

Ⓑ The number of bricks piled up and total height reached when the bricks are 6 cm high.

① Let's summarize the relationships of Ⓐ and Ⓑ in the following tables.

Ⓐ

Length and width of the rectangle

Length (m)	1	2	3	4	5	6	7	
Width (m)	14	13						

Ⓑ

Number of bricks piled up and height

Number of bricks	0	1	2	3	4	5	6	7	
Height (cm)	0	6	12						

The way it changes is different.

When some quantities increase, some decrease.

Hiroto

Nanami

Purpose Which relationship exists between quantities that change together?

② About Ⓐ, if the length increases, how does the width change?

③ About Ⓑ, if the number of bricks increases, how does the height change?

Way to see and think

Try to think what rules exist on how to change.

Summary

Two quantities changing together have a relationship that increases as one quantity increases, and a relationship that decreases as one quantity increases.

2 Proportion

Want to explore

1

There is a ribbon that costs 90 yen per meter. Let's explore the relationship between the length and cost of the ribbon.

① Let's summarize the relationship between the length and cost of the ribbon in the following table.

Length and cost of the ribbon

Ribbon's length (m)	1	2	3					
Ribbon's cost (yen)	90							

② How does the cost change as the ribbon's length increases 1 m, 2 m, …, and so on?

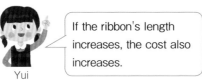

If the ribbon's length increases, the cost also increases.

Yui

For every increase of 1 m, it will increase 90 yen.

Daiki

Purpose What is the relationship between the ribbon's length and cost?

When the ribbon's length is ☐ m and the cost is ○ yen, if ☐ increases then ○ increases accordingly.

Want to find

③ Let's explore how the corresponding cost ○ yen changes when the ribbon's length ☐ m increases 2 times, 3 times, 4 times, …, and so on. Let's fill in each ☐ with a number.

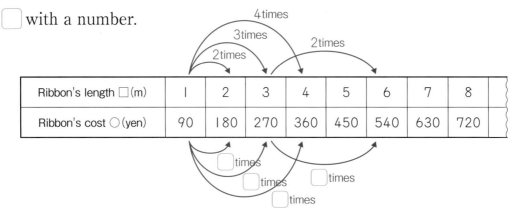

Ribbon's length ☐ (m)	1	2	3	4	5	6	7	8	
Ribbon's cost ○ (yen)	90	180	270	360	450	540	630	720	

If there are two quantities □ and ○ changing together, and □ changes **2 times, 3 times,** ..., so on, and ○ **also** changes **2 times, 3 times,** ..., so on, then ○ **is** called **proportional to** □.

Summary

If the ribbon's length changes 2 times, 3 times, ..., so on, and the cost also changes 2 times, 3 times, ..., so on, then the cost is proportional to the ribbon's length.

④ How much is the cost when the length of the ribbon is 8 m?

8 times

Ribbon's length □ (m)	1	8
Ribbon's cost ○ (yen)	90	?

8 times

Way to see and think

The ribbon's length is 8 times, so the cost is also 8 times.

⑤ Let's represent □ and ○ in a math sentence.

$$\boxed{} \times \square = \bigcirc$$

Ribbon's cost for 1 m   Length   Cost

You can think of it by using tables and math sentences.

⑥ How much is the cost when the length of the ribbon is 12 m? Also, what is the length of the ribbon when the cost is 1800 yen?

Want to confirm

1 Pile up bricks with a height of 6 cm. Let ○ cm be the total height after you pile up □ bricks. Let's answer the following questions.

① Let's summarize the relationship between number of bricks and total height in the table.

Number of bricks and total height

Number of bricks □	1	2	3	4	5	6	7	8
Total height ○ (cm)	6							

② The total height of the bricks is proportional to what?

③ Let's represent the relationship of □ and ○ in a math sentence.

④ What is the total height when 15 bricks are piled up? Also, how many bricks were piled up when the total height is 126 cm?

Want to think

1

As shown in the figure on the right, find the surrounding length of a square that increases its length side by 1 cm. Let's think about the two quantities that change together.

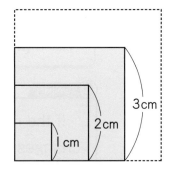

3cm

2cm

1 cm

① With word expressions, let's write a math sentence for the surrounding length, and explore which quantities change together.

$$\boxed{} = \boxed{} \times \boxed{}$$

② What remains unchanged in math sentence ① ?

③ Let's write a math sentence for the surrounding length, by using □ cm as the length of one side and ○ cm as the surrounding length.

④ The following table summarizes the relationship between the length of one side □ cm and the surrounding length ○ cm. How does ○ change when □ is increased 2 times, 3 times, 4 times, ..., so on? Let's fill in the ▢ with numbers.

Way to see and think

2 times is the same as multiplying a number by 2.

× 4

× 3

× 2

Length of one side □ (cm)	1	2	3	4	5	6	7	8	
Surrounding length ○ (cm)	4	8	12	16	20	24	28	32	

× □   × □   × □

⑤ Is the surrounding length proportional to the length of one side? Let's also write the reasons.

⑥ If the surrounding length of a square is 56 cm, then how many cm is the length of one side?

⑦ About the summary table in ④, it was explored how ○ changes when □
is divided by 2, 3, ..., so on. Let's fill in the □ with numbers.

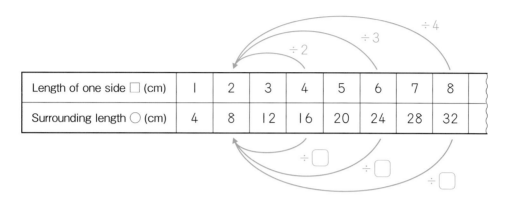

Length of one side □ (cm)	1	2	3	4	5	6	7	8
Surrounding length ○ (cm)	4	8	12	16	20	24	28	32

1 Rectangles with a 5 cm length and 3 cm width are connected as shown in
the following diagram. Let's explore the relationship between the width and the
area as the rectangles are connected.

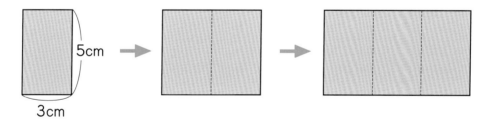

5cm

3cm

① Let's write a math sentence for the area, by using □ cm as the width and
○ cm² as the area.

② Let's summarize the relationship between the width and the area of the
rectangle in the following table.

Width and area of the rectangle

Width □ (cm)	3	6					
Area ○ (cm²)							

③ Is the area of the rectangle proportional to the width? Let's also write the
reasons.

What you can do now

Understanding proportional relationships.

1 On the following Ⓐ~Ⓒ, in which case is ○ proportional to □?

Ⓐ Length □ cm and width ○ cm when a rectangle's surrounding length is 26 cm.

Length □ (cm)	1	2	3	4	5	6	
Width ○ (cm)	12	11	10	9	8	7	

Ⓑ □ number of balls and total cost ○ yen when the cost of one ball is 300 yen.

Number of balls □	1	2	3	4	5	6	
Total cost ○ (yen)	300	600	900	1200	1500	1800	

Ⓒ □ number of candies and total cost ○ yen when the cost of one candy is 8 yen.

Number of candies □	1	2	3	4	5	6	
Total cost ○ (yen)	8	16	24	32	40	48	

Can solve problems on proportion.

2 Water is poured into a water tank so that the depth of water increases 2 cm in 1 minute.

Let's explore the relationship between the time □ (min) to pour water into the water tank

and the depth of water ○ cm.

① Let's summarize the relationship between the time □ (min) to pour water

and the depth of water ○ cm in the table.

Time to pour water and the depth of water

Time to pour water □ (min)	1	2	3	4	5	6	
Depth of water ○ (cm)	2						

② Which is proportional to what?

③ If □ increases by 1, how much does ○ increase?

④ Let's represent the relationship of □ and ○ in a math sentence.

⑤ Let's find the depth of water when the time to pour water is 9 minutes.

⑥ Let's find the time to pour water so that the depth of water is 30 cm.

Usefulness and efficiency of learning

1 Is ○ proportional to □? Let's write a table and confirm. Also, in the case of proportionality, let's represent the relationship between □ and ○ in a math sentence.

Understanding proportional relationships.

① □ number of notebooks and total cost ○ yen when the cost of one notebook is 120 yen.

② Hiroto's age □ years and his 3 year old younger sister's age ○ years when Hiroto's birthday is reached.

③ □ number of practice days and ○ total minutes of practice when the practice is everyday for 30 minutes.

④ Length of one side of a square is □ cm and area ○ cm².

2 As shown in the following diagram, rectangles with length 2 cm and width 6 cm are piled up. Let's explore the relationship between the length and area of the piled rectangles.

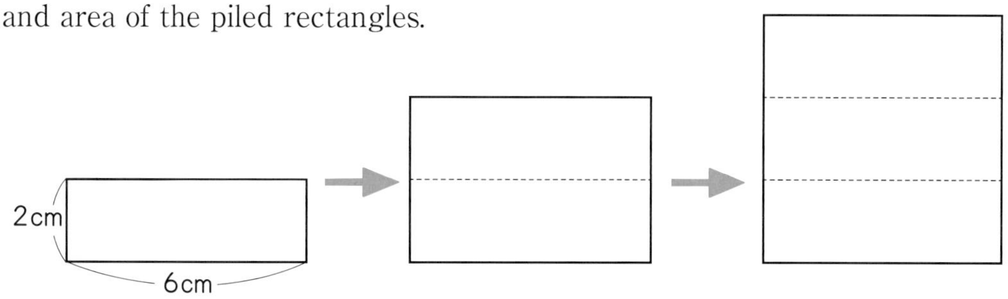

Can solve problems on proportion.

2cm
6cm

① Let's write a math sentence by using length □ cm and area ○ cm².

② Let's summarize the relationship between the length and area of the rectangle in the table.

Length and area of the rectangle

Length □ (cm)	2	4				
Area ○ (cm²)						

③ Is the area of the rectangle proportional to the length? Let's also write the reasons.

How can we make the same amount?

 Problem What should you do to divide into the same amount?

Mean

4 Let's think about how to make each amount the same.

Activity

1

Juice was poured in containers as shown in the diagram on the right. Let's divide so that the amount of juice is the same.

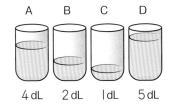

A   B   C   D

4 dL   2 dL   1 dL   5 dL

Why don't you transfer from too many to a few?

I want to pour all juice in one big container and then divide.

Yui   Hiroto

① Let's compare the ideas of the following children.

Yui's idea

Move juice from the one with a large amount to the one with a small amount.

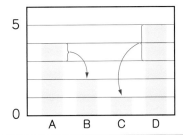

Hiroto's idea

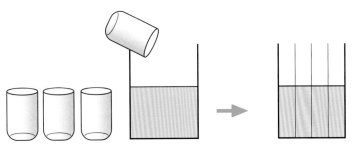

The total amount of juice poured in the 4 containers is

4 + 2 + 1 + 5 = ☐ (dL)

If this juice is equally divided by 4, ☐ ÷ 4 = ☐ (dL)

The same number or measure which was obtained by getting the average of the given numbers or measures is called the **mean**.

Mean = Total ÷ Number of Items

If we try to represent Hiroto's thinking method, it becomes as follows:

$(4 + 2 + 1 + 5) \div 4 = 3$

Total amount of juice Number of containers Mean

Want to try

 The following table shows the number of books that Soshi read from April to August. What is the mean number of books he read in one month?

Number of books he read

Month	April	May	June	July	August
Number of books	4	3	0	2	5

Daiki's idea

$(4 + 3 + 0 + 2 + 5) \div \boxed{} = \boxed{}$

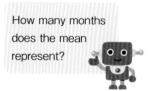

How many months does the mean represent?

Even if something cannot be expressed in a decimal number, such as a number of books, the mean may be expressed in a decimal number.

Words

 【平】 Looking flat and even.

 【均】 Making things equal.

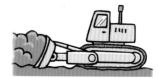

2 A group of 5th graders run in the playground every morning. The following table summarizes how many times Erika, Yuji, and Nozomi ran around the playground last week. Nozomi was sick and rested for one day, so she ran 4 days. Who ran the largest number of times around the playground?

Number of times running around the playground (times)

Day number	Day 1	Day 2	Day 3	Day 4	Day 5
Erika	1	3	4	4	3
Yuji	5	3	7	2	3
Nozomi	7	4	5	4	

① Let's look at the above records by the 3 children and discuss.

If we try to compare in total...

Yui

But Nozomi's total days is 4. Is it good to compare in total even though the number of days is smaller?

Hiroto

If Nozomi had not taken a rest, how many times would she have run that day?

Nanami

② As for the 3 children, what is the mean number of times they ran in one day?

 The following table shows the scores of the 1st and 2nd group of students that took a mathematics test.

Scores of the 1st group in the mathematics test

Name	Yuri	Kota	Saki	Keiya	Rinka
Score (points)	78	65	70	81	90

Scores of the 2nd group in the mathematics test

Name	Sho	Tomomi	Masaya	Chiaki	Itsuki	Fumino
Score (points)	82	63	69	74	88	86

The mean score obtained by each student is called the average score.

① Let's find the average scores of the first and second group of students in the mathematics test.

② Which group has a higher average score?

Words

【If..., ,then... is....】

This is a word that you use when something is assumed or predicted. In mathematics, it is often used when the conditions are altered to get the conclusion.

3 **Let's count the number of steps and find the approximate length of one step.**

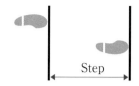

Step

① The following table shows the length Kaito walked with 10 steps. How many meters is one step? Let's find it out by rounding off to the nearest hundredth.

> The steps are not always the same, so you need to measure several times to find the mean.

Daiki

Length Kaito walked with 10 steps

Number of times	1st Time	2nd Time	3rd Time
Length walked with 10 steps (m)	5.17	5.13	5.18

Mean number of the 3 times: $(5.17 + 5.13 + 5.18) \div 3 = $ ☐

One step: $5.16 \div 10 = $ ☐ Approximately ☐ m

② Kaito counted 1731 steps from home to school. Using the result from ①, can you think how many meters is the approximate distance from home to school? Let's find it out by rounding off to the nearest whole number.

③ Let's examine the length of your own step, and try to explore the approximate length of various places.

How to find the mean when measuring

4 **Megumi examined, during a science experiment, the time it takes a pendulum to go back and forth (period) 10 times. Let's find the mean number for 1 period on the pendulum.**

Time it takes a pendulum to go back and forth

Number of times	1st Time	2nd Time	3rd Time	4th Time
Time for 10 periods (sec)	15	14	23	13
Time for 1 period (sec)	1.5	1.4	2.3	1.3

> There is a value far from other values.

Yui

① Let's look at the above table, and discuss how to find the answer.

② Nanami thought as follows. Let's explain Nanami's idea.

Nanami's idea

Since I think the 3rd time failed, the mean is found by excluding this value.

As for this experiment, the mean is found using the 1st time, 2nd time, and 4th time. Therefore, divided in ☐ parts.

(☐ + ☐ + ☐) ÷ ☐ = ☐

First of all, let's explore why the distant values came out.

If you find the mean with all the values,
(1.5+1.4+2.3+1.3)÷4=1.625,
then it is about 1.6 seconds.

Hiroto

When there is a large distant value, the result may be influenced, so in that case it is better to calculate by excluding that value.

 3 The following table shows the records when Arata ran 50 m. How many seconds is the mean number for Arata's record?

Arata's records for a 50 m run

Number of times	1st Time	2nd Time	3rd Time	4th Time	5th Time
Record (sec)	9.5	9.1	9.4	12.6	9.2

The 4th time record is a large distant value from the other values.

Daiki

5 As for chickens Ⓐ and Ⓑ, which laid heavier eggs?

Ⓐ

A 56g B 58g C 56g

D 61g E 54g F 57g

Ⓑ

G 57g H 53g J 60g K 58g

L 56g M 53g N 55g

① Let's find and compare the mean weights.

② Yui thought the following to find the mean weight of eggs laid by chicken Ⓐ. What was the improvement done by Yui?

Yui's idea

Based on the lightest egg, 54 g, I thought how heavy the other eggs were.

56 58 56 61 54 57 (g)

↓   ↓   ↓   ↓   ↓   ↓

2   4   2   7   0   3 (g)

(Reference)

I found this mean number by,

$(2 + 4 + 2 + 7 + 0 + 3) ÷ 6 = 3$

If this mean number is added to the weight I used as a reference, I can find the mean weight of the eggs.

$54 + 3 = 57$

The mean weight of the eggs laid by Ⓐ is 57 g.

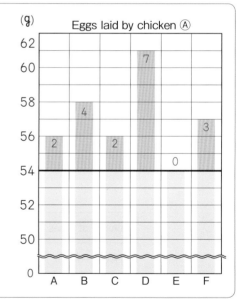

(g) Eggs laid by chicken Ⓐ

③ Thinking in the same way, let's find the mean weight for the eggs laid by chicken Ⓑ.

Even for things that you can't distribute evenly, if you know the number or amount, you can calculate the mean.

④ The total weight of eggs laid by chicken Ⓑ was 728 g. Can you calculate how many eggs were laid?

What you can do now

Can find a mean number.

1 The following table shows the number of empty cans picked by Marina in 5 days.

Number of empty cans picked

Day number	1st Day	2nd Day	3rd Day	4th Day	5th Day
Number of empty cans	6	7	5	9	8

① What is the mean number of cans picked in one day?

② If Marina picks up empty cans in the same way for 15 days, can you calculate the total number of empty cans picked?

Can use a mean number.

2 Toshiki walked 1850 steps from his house to the library. If Toshiki's step is about 0.54 m, can you calculate how many meters is the distance from Toshiki's house to the library?

Can find the mean when measuring.

3 The following table shows the records for Sayaka's softball throws.

Records for softball throws

Number of times	1st Time	2nd Time	3rd Time	4th Time
Records (m)	24	22	12	26

Can you calculate how many meters is the mean for Sayaka's records?

Can solve problems using mean numbers.

4 1st and 2nd group of 5th grade went to a potato field. The number of people in each group and the total number of potatoes are shown in the table on the right.

Number of harvested potatoes

	Number of people	Total number
1st Group	24	84
2nd Group	30	102

Can you say which group harvested the most number of potatoes? What is the mean number of potatoes harvested by 1 person? Let's compare.

Supplementary problems p.148

Usefulness and efficiency of learning

1 The following table examines the number of tomatoes harvested by pot planting from Monday to Friday. What is the mean number of tomatoes harvested in one day?

☐ Can find a mean number.

Number of harvested tomatoes by pot planting

Day of the week	Monday	Tuesday	Wednesday	Thursday	Friday
Number of tomatoes	6	3	2	0	8

2 Kazuya's objective is to read a mean number of 25 pages a day. The mean number for 6 days (from Sunday to Friday) was 23 pages. How many pages should Kazuya read on Saturday, so that in a 7 days period (from Sunday to Saturday), he achieves the mean number objective of 25 pages a day?

☐ Can use a mean number.

3 The following table shows the records for Yuki's long jumps.

☐ Can find the mean when measuring.

Records for long jump

Number of times	1st Time	2nd Time	3rd Time	4th Time	5th Time	6th Time
Record	2m45cm	2m32cm	2m44cm	2m48cm	1m25cm	2m36cm

Can you think how many m and cm is Yuki's mean record?

4 There are 6 onions inside a box.
When weighed, it became as shown on the right.
Let's fill in each ☐ with a number.

162g, 157g, 159g,
161g, 160g, 158g

☐ Can solve problems using mean numbers.

The mean weight of the onions inside the box was found as follows.

Based on the smallest weight, 157 g, we thought how much was the weight of the other onions.

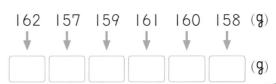

162 157 159 161 160 158 (g)
↓   ↓   ↓   ↓   ↓   ↓
☐   ☐   ☐   ☐   ☐   ☐   (g)

We found the mean number by,

(☐ + ☐ + ☐ + ☐ + ☐ + ☐) ÷ 6 = ☐

so the mean weight of the onions was,

157 + ☐ = ☐   ☐ g

Let's **deepen.**

While measuring, can I also use mean numbers when resulting between two scales?

Daiki

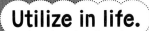

Measuring an amount smaller than given scale?

Want to know

When the temperature in the classroom was measured with a thermometer, it came between two scales as shown in the diagram on the right. Then, after 9 children took the temperature separately, it became like the following table. Can you say how many degrees was the temperature of the classroom?

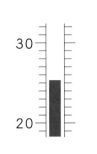

Temperature taken separately

Name	Sho	Takeshi	Noriko	Tomomi	Jun	Mika	Daiki	Nana	Kota
Temperature (℃)	25.2	25.1	25.3	25.2	25.2	25.1	25.4	25.2	25.1

① Let's find the mean temperature taken by the 9 children.

② Let's try to discuss whether the temperature found in ① can be said to be the correct temperature.

When measuring with a tool and resulting between two scales, the mean of the results taken by several people can be the correct result.

Want to deepen

When the length of one side of an equilateral triangle board was measured, according to the position of 0 on the left of the ruler, the right side came between two scales. Then, after 9 children made the measurement separately, it became like the following table. Can you say how many mm is the mean length of one side of the equilateral triangle board?

Length of one side measured separately

Name	Sho	Takeshi	Noriko	Tomomi	Jun	Mika	Daiki	Nana	Kota
Length (mm)	87.1	87.3	87.1	87.2	87.2	87.3	87.1	87.4	87.1

How crowded ?

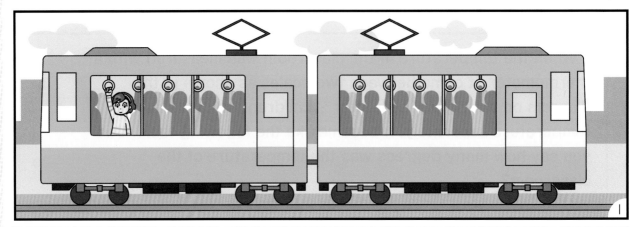

1

2

Today's train was very crowded.

The train that I rode was more crowded.

You can't compare because the size of the train and the number of people on board were different.

3

How can you compare crowdedness of people?

50

5 Measure per Unit Quantity(1)

Let's think about the way to change the unit for comparing.

Activity

Want to explore How to compare crowdedness

1 Children are standing on mats. Let's explore which of the following Ⓐ, Ⓑ, and Ⓒ is the most crowded?

Number of children standing on mats

	Number of mats	Number of children
Ⓐ	2	12
Ⓑ	3	12
Ⓒ	3	15

Ⓐ 12 children standing on 2 mats.

Ⓑ 12 children standing on 3 mats.

 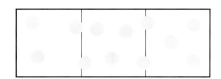

Ⓒ 15 children standing on 3 mats.

 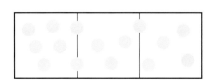

I wonder if the one with the most number of people Ⓒ is the most crowded.

Daiki

Since the number of mats is different...

Nanami

Purpose What should we do to compare crowdedness of people?

① Between Ⓐ and Ⓑ, which one is more crowded?

Ⓐ   12 children standing on 2 mats.   Ⓑ   12 children standing on 3 mats.

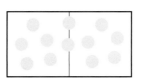 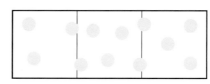

② Between Ⓑ and Ⓒ, which one is more crowded?

Ⓒ   15 children standing on 3 mats.

Way to see and think

A comparison can be made If you have either the same number of people or the same number of mats.

Want to compare

③ Between Ⓐ and Ⓒ, which one is more crowded? Let's explain and compare the ideas of the following children.

 Nanami's idea

Compare how many children are standing on one mat.

Ⓐ $12 \div 2 = 6$

     6 children standing on one mat.

Ⓒ $15 \div 3 = 5$

     5 children standing on one mat.

Ⓐ   Ⓒ

Ⓐ is more crowded because there are more children standing on one mat.

Way to see and think

Comparing the number of children standing on one mat using the mean number idea.

 Yui's idea

Compare how much of a mat is occupied by one child.

Ⓐ $2 \div 12 = 0.166\cdots$ About 0.17 mat per child.

Ⓒ $3 \div 15 = 0.2$ About 0.2 mat per child.

Ⓐ is more crowded because the area occupied by one child is smaller.

Way to see and think

Comparing the area occupied by one child.

Hiroto's idea

Compare the number of children standing on 6 mats.

Ⓐ 12 × 3 = 36   36 children standing on 6 mats.

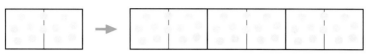

Ⓒ 15 × 2 = 30   30 children standing on 6 mats.

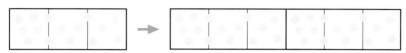

When 6 mats are aligned, Ⓐ is more crowded.

Way to see and think

Comparing the number of children standing on aligned mats using the proportion idea.

④ The area of one mat is 1 m². As for Ⓐ, Ⓑ, and Ⓒ, how many children are standing per square meter?

Ⓐ 12 ÷ 2 = ☐

Ⓑ 12 ÷ 3 = ☐

Ⓒ 15 ÷ 3 = ☐

It becomes more crowded as the number of people per square meter increases.

Daiki

Number of children   Area (m²)   Number of children per square meter

🌼 **Summary**

Crowdedness can be compared by the number of people per square meter or area occupied per person.

Usually, crowdedness of people is compared aligning the same unit, such as m² or km².

Want to confirm

1 ▶ There are 10 children playing in a 8 m² sandbox. Next to it, there are 13 children playing in a 10 m² sandbox. Which sandbox is more crowded?

2 ▶ There is a train with 7 cars carrying 1260 passengers while another carries 1850 passengers in 10 cars. Which train is more crowded?

2

The table on the right represents the population and area of East City and West Town. Let's find the number of people per square kilometer and compare how crowded is each place.

Population and area

	Population (people)	Area (km²)
East City	273600	72
West Town	22100	17

Way to see and think

Population density is considered using the same area unit.

The number of people per square kilometer is called **population density.** The crowdedness of people living in a country or prefecture is represented by the population density.

3 Choose 3 prefectures and calculate the population density of each prefecture. Let's round off the tenths place and express the answer in whole numbers.

Let's try to explore your own town.

Population in 2015.

Hokkaido
83424km²
5,381,733 people

Aomori Prefecture
9646km²
1,308,265 people

Niigata Prefecture
12584km²
2,304,264 people

Hiroshima Prefecture
8479km²
2,843,990 people

Osaka
1905km²
8,839,469 people

Tokyo
2191km²
13,515,271 people

Fukuoka Prefecture
4986km²
5,101,556 people

Shizuoka Prefecture
7777km²
3,700,305 people

Kumamoto Prefecture
7409km²
1,786,170 people

Kagawa Prefecture
1877km²
976,263 people

Kagoshima Prefecture
9187km²
1,648,177 people

Kochi Prefecture
7104km²
728,276 people

Okinawa Prefecture
2281km²
1,433,566 people

Activity

3

A wire is 8 m long and weighs 480 g. Let's think about how to find the weight of the wire per meter.

Can I think using a diagram?

Daiki

Thinking with a table...

Nanami

Want to compare

① Using diagrams and tables, a math sentence to find the weight per meter was considered. Let's explain and compare the ideas of the following children.

Daiki's idea

I considered the weight per meter as ☐ g.

Way to see and think

Using the idea of measurement of one unit × how many units = total measurement.

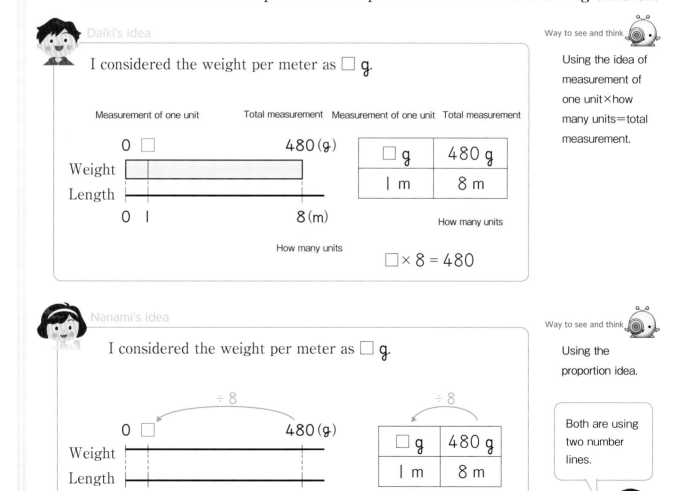

Nanami's idea

I considered the weight per meter as ☐ g.

Way to see and think

Using the proportion idea.

Both are using two number lines.

Hiroto

② How many grams is the weight per meter?

③ How many grams is the weight of a 15 m long wire? Let's think using a diagram and table.

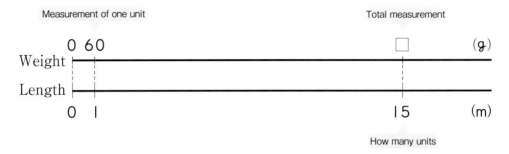

Measurement of one unit	Total measurement
60g	□ g
1m	15m

How many units

④ This wire was cut and weighed 300 g. How many meters is the length of the piece of wire?

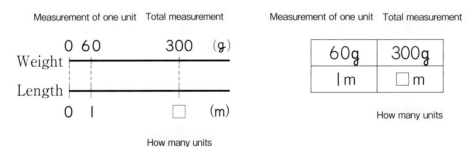

Measurement of one unit	Total measurement
60g	300g
1m	□ m

How many units

 Population density, weight per meter, etc. are called **measure per unit quantities**.

Want to confirm

 A wire is 15 m long and weighs 600 g. How many grams is the weight per meter of this wire?

4 At school, potatoes were harvested as follows. A total weight of 43.2 kg was harvested from a 6 m² field, and 62.1kg were harvested from a 9 m² field. Can you say which field had a better harvest? Let's compare by the weight of harvested potatoes per square meter.

Measure per unit quantity   Total measurement

```
         0   □            43.2              (kg)
Weight ├───┼──────────────┼──────────────────
Area   ├───┼──────────────┼──────────────────
         0   1             6                 (m²)
```
How many units

Measure per unit quantity   Total measurement

□ kg	43.2kg
1 m²	6m²

How many units

```
         0   □            62.1              (kg)
Weight ├───┼──────────────┼──────────────────
Area   ├───┼──────────────┼──────────────────
         0   1             9                 (m²)
```

□ kg	62.1kg
1 m²	9m²

5 There are 10 notebooks that cost 1200 yen in total and 8 notebooks that cost 1040 yen in total. Can you say which notebook is more expensive?

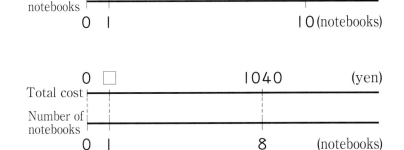

```
            0   □                  1200    (yen)
Total cost ├──┼───────────────────┼────────
Number of  ├──┼───────────────────┼────────
notebooks   0   1                  10 (notebooks)
```

□ yen	1200 yen
1 notebook	10 notebooks

```
            0   □            1040         (yen)
Total cost ├──┼─────────────┼────────────
Number of  ├──┼─────────────┼────────────
notebooks   0   1           8  (notebooks)
```

□ yen	1040 yen
1 notebook	8 notebooks

6 There is a car that runs 720 km with 45 L of gasoline. Let's answer the following questions about this car.

① Let's find the distance that runs per liter of gasoline.

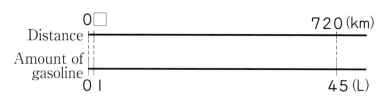

Measure per unit quantity   Total measurement

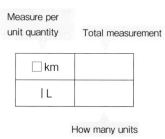

How many units

② How many **km** will the car run with 32 L of gasoline?

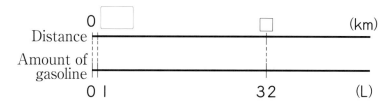

③ To run 1024 km, how many liters of gasoline are needed?

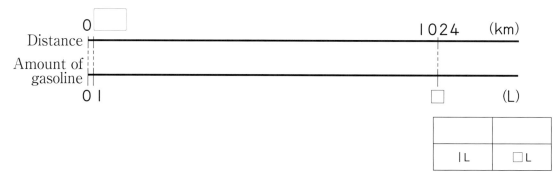

7 There is a car that runs 390 km with 30 L of gasoline. Let's answer the following questions about this car.

① Let's find the distance that runs per liter of gasoline.

② How many kilometers will the car run with 47 L of gasoline?

③ To run 182 km, how many liters of gasoline are needed?

What you can do now

☐ **Can compare crowdedness.**

1 From the following Ⓐ and Ⓑ, which is more crowded?

① Ⓐ 5 children standing on 2 mats.

Ⓑ 6 children standing on 3 mats.

② Ⓐ 15 children standing on 4 mats.

Ⓑ 11 children standing on 3 mats.

③ Ⓐ 17 children standing on 6 mats.

Ⓑ 14 children standing on 3 mats.

☐ **Can find the population density.**

2 The population of Haruka's city is about 39000 people with an area of about 50 km².

The population of the neighboring city is about 45000 people with an area of about 65 km².

① Let's find the population density of Haruka's city.

② Which city has the highest population density? Haruka's city or the neighboring city?

☐ **Can find and compare measure per unit quantities.**

3 Let's answer the following questions.

① There are 12 colored pencils for 600 yen and 8 colored pencils for 440 yen. Can you say which colored pencils is more expensive?

Let's compare by the price per pencil.

② A wire is 5 m long and weighs 370 g. Let's think about this wire.

ⓐ Let's find the weight per meter.

ⓑ How many grams is the weight of a 10 m long wire?

ⓒ A wire of the same type was measured and weighed 555 g.

How many meters is the length of this wire?

Supplementary problems ●●●●●●●● ➤ p.149

Usefulness and efficiency of learning

1 From the following Ⓐ and Ⓑ, which train is more crowded?

Can compare crowdedness.

① Ⓐ A train with 6 cars and 1080 passengers.

　Ⓑ A train with 8 cars and 1640 passengers.

② Ⓐ A train with 10 cars and 1840 passengers.

　Ⓑ A train with 15 cars and 2700 passengers.

2 The following table shows the population and area of two cities and one town. Let's answer the following questions.

Can find the population density.

	Population (people)	Area (km²)
North City	267200	70
South City	185000	47
East Town	58000	14

① Let's find the population density of North City and South City. Round off the tenths place and express the answer in whole numbers.

② Let's arrange the population density in descending order.

3 Let's answer the following questions.

Can find and compare measure per unit quantities.

① From the orange field located on the west side of Yuma's house, 432 kg of oranges were harvested in an area of 180 m². And from the field at the south side, 247 kg of oranges were harvested in an area of 130 m². Let's answer the following questions about this orange fields.

ⓐ How many kilograms of oranges per square meter were harvested at the west side field?

ⓑ Which field had a better harvest of oranges?

② Let's answer the following questions about a car that runs 360 km with 15 L of gasoline.

ⓐ How many liters of gasoline will this car use to run for 840 km?

ⓑ How many kilometers can this car run with 27 L of gasoline?

Let's deepen.

Is there any comparison by using measure per unit quantity in my surroundings?

Daiki

Deepen.

Let's look at the environment by a measure per unit quantity.

Want to know

As I investigated global warming, I heard that one of the causes was the increase in the amount of carbon dioxide in the air. The following table shows the examined information about how much it increased in Japan.

Year	Amount of carbon dioxide emissions (ten thousand kg)	Population (ten thousand people)	Amount of carbon dioxide emissions per person (kg)
1990	116200000	12361	
1995	124800000	12557	
2000	128000000	12693	
2005	131100000	12777	
2010	121700000	12806	
2015	122700000	12709	

① How does the amount per person increase? Let's try to think and represent using a bar graph or line graph.

Want to deepen

The following graph shows the amount of carbon dioxide emissions per person in various countries. What can you say? Let's try to discuss.

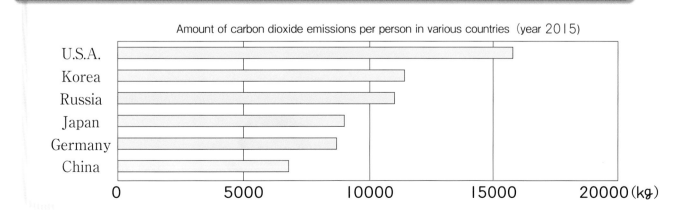

Amount of carbon dioxide emissions per person in various countries (year 2015)

Reflect

Connect

Problem

Let's try to solve various problems.
---

① **Divide a string, that is 12m long, into 3 equal parts. How many meters is the length of one part?**

(Math sentence) $12 ÷ 3 = 4$

answer   4m

Representing in a table...

□m	12m
1 string	3 strings

The same as measure per unit quantity.

1 string unit	=	per string

② **The area for one sheet of origami is 225cm². How many cm² is the area covered by 4 sheets of origami?**

(Math sentence) $225 × 4 = 900$

Answer   900cm²

225cm²	□cm²
1 sheet	4 sheets

Similar to the problem on page 56.

All of these were learned in the 3rd grade, however, it is similar to the measure per unit quantity that we are learning now.

Nanami

Daiki

1 string unit	=	per string
1 sheet unit	=	per sheet

Don't you?

③ There is 1L 5dL of juice. If every time you drink 3dL, how many times can you drink?

1L 5dL = 15dL

(Math sentence)   15 ÷ 3 = 5

Answer   5 times

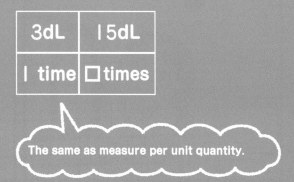

3dL	15dL
1 time	☐ times

The same as measure per unit quantity.

If the units are aligned with L ...

1L 5dL = 1.5 L
3dL = 0.3 L

1.5 ÷ 0.3 = ?

Summary

So far, multiplication and division have used the idea of measure per unit quantity.

The multiplications and divisions learned so far were related to the measure per unit quantity.

Yui

Want to connect

Sometimes the divisor is a decimal number. Can it be calculated?

Hiroto

How much is the ribbon?

Problem What kind of operation do you use to find the price for 2.4 m of ribbon?

6

Multiplication of Decimal Numbers

Let's think about how to calculate and use the rules of operations.

l m

80 yen

2.4m

[ ] yen

1 Operation of whole number × decimal number

Want to solve Multiplication of decimal numbers

1
You buy a ribbon that costs 80 yen per meter. How much should you pay for 2.4 m of ribbon?

① Let's consider a math sentence to find the price by using a diagram and table.

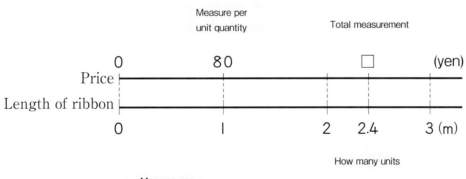

Measure per unit quantity　　Total measurement

	0	80				(yen)
Price				□		

Length of ribbon

| 0 | l | 2 | 2.4 | 3 (m) |

How many units

Measure per unit quantity　　Total measurement

80 yen	□ yen
l m	2.4m

How many units

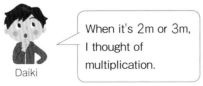

When it's 2m or 3m, I thought of multiplication.

Daiki

If the price is proportional to the length of the ribbon ...

Yui

Purpose As for the price for 2.4 m of ribbon, is it correct to think the same as in multiplication of whole numbers?

② Let's explain the ideas of the following children.

Diki's idea

Measure per unit quantity × How many units = Total measurement,

Way to see and think

When the "how many units" is a decimal number, can we do the same as with whole numbers?

	Price for 1m		Length		Price
When the length of the ribbon is 2 m	80	×	2	=	160
3 m	80	×	3	=	240
2.4 m	80	×	☐	=	☐

Yui's idea

The length of the ribbon is 2.4 times, the price will be 2.4 times as well.

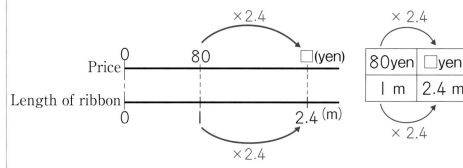

Way to see and think

If the length of the ribbon is 2.4 times, will the price be 2.4 times as well?

Summary

Even when the "how many units" is a decimal number, such as the length of the ribbon, the total measurement can be calculated with a multiplication in the same way as with whole numbers.

③ About how many yen is the price?

④ Let's think about how to calculate 80 × 2.4.

Purpose What should we do to calculate whole number × decimal number?

⑤ Let's compare the calculation methods of the following children.

Hiroto's idea

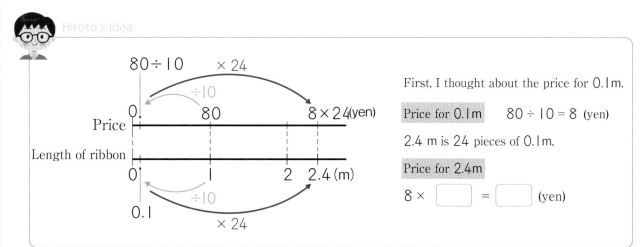

First, I thought about the price for 0.1 m.

Price for 0.1 m $80 ÷ 10 = 8$ (yen)

2.4 m is 24 pieces of 0.1 m,

Price for 2.4m

$8 × \boxed{} = \boxed{}$ (yen)

Nanami's idea

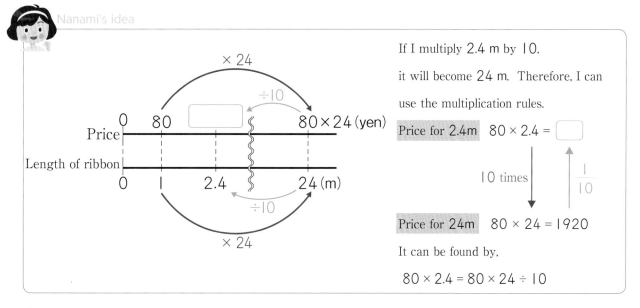

If I multiply 2.4 m by 10,

it will become 24 m. Therefore, I can

use the multiplication rules.

Price for 2.4m $80 × 2.4 = \boxed{}$

10 times $\frac{1}{10}$

Price for 24m $80 × 24 = 1920$

It can be found by,

$80 × 2.4 = 80 × 24 ÷ 10$

Summary

   As for the calculation of whole number × decimal mumber, if the decimal number is changed to a whole number then the answer can be found.

1 ▶   How many yen is the cost for 2.3 m of ribbon that has a price of 90 yen per meter?

   Let's find out using Hiroto's and Nanami's ideas from above.

2 Let's think about how to calculate 80 x 2.4 in vertical form.

```
      8 0
  ×   2 . 4
```

Diaki: Can I use the method for decimal number × whole number?

Is the position of the decimal point the same as in decimal number × whole number?

 Nanami

Purpose What should we do to calculate whole number × decimal number in vertical form?

① Let's explain how to calculate in vertical form.

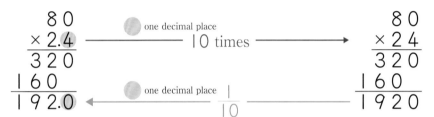

```
    8 0              one decimal place          8 0
  × 2.4   ————— 10 times ————→            × 2 4
  ─────                                    ─────
  3 2 0                                    3 2 0
  1 6 0                                    1 6 0
  ─────                                    ─────
1 9 2.0   ←——— one decimal place  1/10 ——— 1 9 2 0
```

Who had the same idea in exercise ⑤ on page 67?

Summary

Whole number × decimal number calculation is performed in the same way as whole numbers calculation, assuming there is no decimal point. The number of digits after the decimal point of the product is the same as the number of decimal places in the decimal number.

 There is a wire that weighs 3 g per meter.

What is the weight for 2.5 m of wire?

① Let's write a math expression. [   ]

② Let's calculate in vertical form.

```
          3
  ×   2 . 5
```

 Let's solve the following calculations in vertical form.

① 60 × 4.7   ② 50 × 3.9   ③ 7 × 1.6

④ 6 × 2.7   ⑤ 24 × 3.3   ⑥ 13 × 2.8

Want to solve

Activity

1

I dL of paint was used to paint 2.1 m² of a wall. How many square meters can be painted with 2.3 dL of paint?

① Let's think, using a diagram and a table, a math expression to find the covered area.

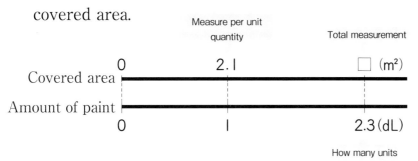

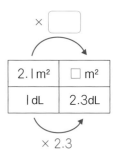

② Let's write a math expression.

Covered area with I dL   Amount of Paint

Want to explain

③ Let's explain the calculation methods of the following children.

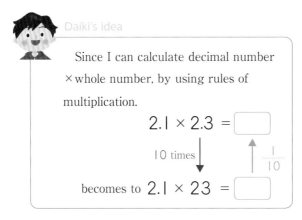

Daiki's idea

Since I can calculate decimal number ×whole number, by using rules of multiplication.

2.1 × 2.3 = ☐

10 times ↓   ↑ $\frac{1}{10}$

becomes to 2.1 × 23 = ☐

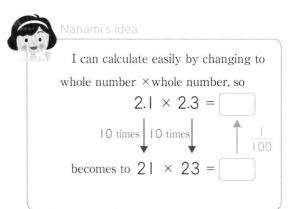

Nanami's idea

I can calculate easily by changing to whole number ×whole number, so

2.1 × 2.3 = ☐

10 times ↓ 10 times ↓   ↑ $\frac{1}{100}$

becomes to 21 × 23 = ☐

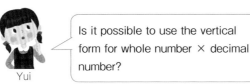

Is it possible to use the vertical form for whole number × decimal number?

Yui

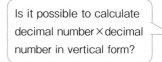

Is it possible to calculate decimal number×decimal number in vertical form?

Hiroto

Purpose Is the calculation of decimal number × decimal number in vertical form the same as before?

④ Let's explain how to calculate 2.1×2.3 in vertical form.

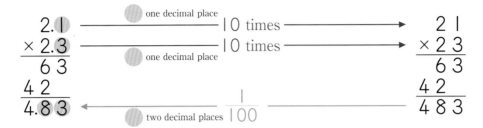

 Let's think about how to calculate 5.26×4.8 in vertical form.

$$\begin{array}{r} 5.2\ 6 \\ \times\quad 4.8 \\ \hline \end{array}$$

two decimal places

$$\begin{array}{r} 5.26 \\ \times\quad 4.8 \\ \hline 4\ 2\ 0\ 8 \\ 2\ 1\ 0\ 4 \\ \hline 2\ 5.2\ 4\ 8 \end{array}$$

100 times
10 times
one decimal place
three decimal places
1000

$$\begin{array}{r} 5\ 2\ 6 \\ \times\quad 4\ 8 \\ \hline 4\ 2\ 0\ 8 \\ 2\ 1\ 0\ 4 \\ \hline 2\ 5\ 2\ 4\ 8 \end{array}$$

Summary Multiplication algorithm of decimal numbers in vertical form

① Calculate in the same way as the calculation of whole numbers, assuming that there is no decimal point.

② Place the decimal point of the product such that the number of decimal places is the same as the sum of the decimal places of the numbers multiplied.

$$\begin{array}{r} 2.1 \cdots \boxed{1} \text{ decimal place} \\ \times\ 2.3 \cdots \boxed{1} \text{ decimal place} \\ \hline 6\ 3 \\ 4\ 2 \\ \hline 4.8\ 3 \cdots ② \text{ decimal places} \end{array}$$

$$\begin{array}{r} 5.26 \cdots \boxed{2} \text{ decimal places} \\ \times\quad 4.8 \cdots \boxed{1} \text{ decimal place} \\ \hline 4\ 2\ 0\ 8 \\ 2\ 1\ 0\ 4 \\ \hline 2\ 5.2\ 4\ 8 \cdots ③ \text{ decimal places} \end{array}$$

Think of 4.36 as 4, and 7.5 as 8 ...

2 Let's think about how to calculate 4.36 × 7.5.

Hiroto

① What is the approximate answer?

② Let's think about it in vertical form.

```
      4 . 3 6  ─── [   ] times ───→    4   3   6
   ×      7 . 5  ─── [   ] times ───→  ×     7   5
      2 1 8 0                         2   1   8   0
   3 0 5 2                          3 0 5 2
   3 2 . 7 0 0  ←── [   ] ──        3 2 7 0 0
```

0 can be erased.

In the following calculations, let's place the decimal point of the product.

①
```
     5.6
   × 4.3
   1 6 8
   2 2 4
   2 4 0 8
```

②
```
     3.2 7
   ×   1.2
     6 5 4
   3 2 7
   3 9 2 4
```

③
```
     1.4 8
   ×   2.5
     7 4 0
   2 9 6
   3 7 0 0
```

Let's solve the following calculations in vertical form.

① 0.2 × 1.6

```
      0 . 2
   ×  1 . 6
```

② 0.4 × 0.35

```
         0 . 4
   ×  0 . 3 5
```

Let's be careful about 0.

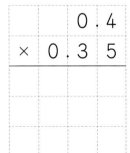

4 Let's solve the following calculation in vertical form.

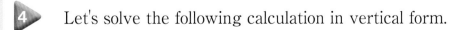

① 1.2 × 2.4   ② 6.4 × 3.5   ③ 3.14 × 2.6

④ 1.4 × 4.87   ⑤ 8.2 × 2.25   ⑥ 0.3 × 1.7

⑦ 0.5 × 2.3   ⑧ 0.43 × 2.1   ⑨ 1.15 × 0.6

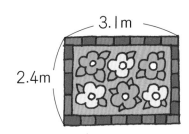

3

How many square meters is the area of a rectangular flowerbed that has a length of 2.4 m and a width of 3.1 m?

Nanami: Can I think using the figure?

Daiki: Can I use the formula for area?

Purpose Even when the length and width are decimal numbers, can we use the formula for area?

Nanami's idea

One square meter is divided into 10 squares for each side, therefore 0.01 m² represents $\frac{1}{100}$ of 1 m². I thought how many of these parts is each side of the flowerbed.

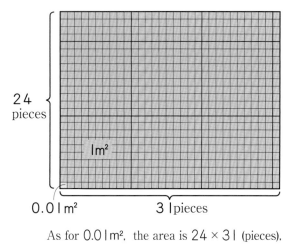

24 pieces

1 m²

0.01 m²   31 pieces

As for 0.01 m², the area is 24×31 (pieces), therefore the answer is ☐ m².

Daiki's idea

If I apply the formula for area,

$$2.4 \times 3.1$$

and consider it in vertical form,

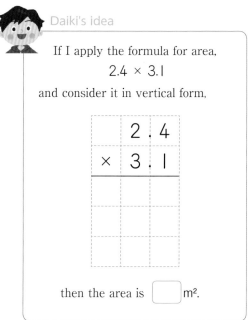

then the area is ☐ m².

Summary

The area can be found applying the formula even if the lengths of the sides are decimal numbers.

How many square meters is the area of a rectangular flowerbed that has a length of 0.6 m and a width of 2.5 m?

4 There is a metal bar that weighs 3.1 kg per meter. Let's find the weight of two bars with a length of 1.2 m and 0.8 m.

The weight of the bar that is 1.2 m is 1.2 times the weight of a bar that is 1 m.

Yui

0.8 m is shorter than 1 m.

Hiroto

Purpose What is the size of the product when the multiplier is a decimal number larger than 1 and smaller than 1?

① Let's write a math expression to find the weight of the metal bar that is 1.2 m.

② Let's write a math expression to find the weight of the metal bar that is 0.8 m.

③ Let's explain about the size of the product and size of the multiplicand using the following number line.

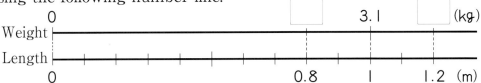

Summary

When the multiplier is a decimal number larger than 1, the product is larger than the multiplicand.

When the multiplier is a decimal number smaller than 1, the product is smaller than the multiplicand.

When the multiplier is 1, the product is the same as the multiplicand.

6 There is a cooking oil that weighs 0.9 kg per liter. Let's find what is the weight of 0.4 L of this cooking oil.

① Let's write a math expression.

Since 0.4 is smaller than 1, then the weight of 0.4 L is less than 0.9 kg.

Nanami

② Let's solve in vertical form.

7 Let's place the decimal point of the products and compare products and multiplicands.

①
```
    2 5        2 5
×    6      × 0.6
  1 5 0      1 5 0
```

②
```
      2.5       0.2 5
×      6      ×   0.6
  1 5 0         1 5 0
```

③
```
    1.6         1.6
× 0.2 4      × 2.4
  3 8 4       3 8 4
```

④
```
  0.0 7       0.0 7
×   0.2      ×     2
    1 4         1 4
```

8 Using six of the following 8 cards, let's make a math expression that can be calculated in vertical form as shown on the right.

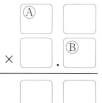

```
 0   1   2   3   4   5   6   7
```

In addition, you cannot place 0 in Ⓐ and Ⓑ, and each card can only be used once. Let's answer the following questions.

① In the case that the multiplicand is 13, let's find the numbers that apply to the above calculation.

② The multiplier will never be 0.1 nor 0.5, for any number you use as the multiplicand. Let's explain the reason.

Want to explore

1

In the calculation of whole numbers, the following rules of operations are valid.

Let's explore whether decimal numbers also hold these rules of operations.

Ⓐ　■ × ▲ = ▲ × ■

Ⓑ　(■ × ▲) × ● = ■ × (▲ × ●)

Ⓒ　(■ + ▲) × ● = ■ × ● + ▲ × ●

Ⓓ　(■ − ▲) × ● = ■ × ● − ▲ × ●

Nanami: When two whole numbers are multiplied, the product is the same even if the multiplicand and the multiplier are interchanged.

Can I do the same with decimal numbers?

Daiki

Purpose Are the rules of operations also valid for decimal numbers?

Want to explain

① Let's explain whether the rule of operation Ⓐ holds by looking at the figure on the right.

3.6 × 2.4 = ☐

2.4 × 3.6 = ☐

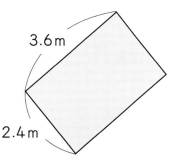

3.6 m

2.4 m

Way to see and think

Which side do you think is it the length?

② Let's explain whether the rule of operation Ⓑ holds by looking at the figure on the right.

(3.6 × 2.5) × 4 = ☐

3.6 × (2.5 × 4) = ☐

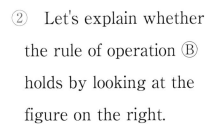

2.5m　2.5m　2.5m　2.5m

3.6m

☐

Way to see and think

The math expression changes depending on whether you see four small rectangles or one large rectangle.

③ Let's explain whether the rule of operation Ⓒ holds by looking at the following figure.

$(1 + 0.4) \times 3 =$ ⬚

$1 \times 3 + 0.4 \times 3 =$ ⬚

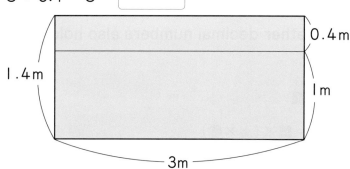

0.4m

1.4m

1m

3m

④ Let's explain whether the rule of operation Ⓓ holds by looking at the following figure.

$(2 - 0.2) \times 3 =$ ⬚

$2 \times 3 - 0.2 \times 3 =$ ⬚

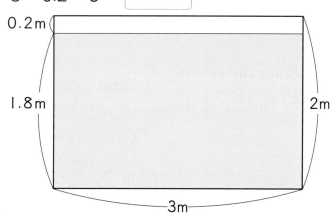

0.2m

1.8m

2m

3m

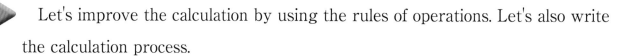

🌼 Summary

Even for decimal numbers, the rules of operations that are valid for whole numbers, are also valid.

Want to try

▶1 Let's improve the calculation by using the rules of operations. Let's also write the calculation process.

① $6.9 \times 4 \times 2.5$

② $0.5 \times 4.3 \times 4$

③ $3.8 \times 4.8 + 3.8 \times 5.2$

④ $1.3 \times 12.9 - 1.3 \times 2.9$

What you can do now

Understanding how to multiply decimal numbers.

1 Let's summarize how to multiply decimal numbers.

To calculate 2.3×1.6, first multiply 2.3 by ☐ and multiply 1.6 by ☐.

Then, calculate ☐ $\times$ ☐. Finally, the answer 368 is multiplied by ☐.

Therefore, $2.3 \times 1.6 =$ ☐.

Can calculate whole number×decimal number and decimal number×decimal number in vertical form.

2 Let's solve the following calculations in vertical form.

① 50×4.3 ② 6×1.8 ③ 26×3.2 ④ 4×1.9

⑤ 31×5.2 ⑥ 62×0.7 ⑦ 0.6×0.8 ⑧ 3.3×0.9

⑨ 1.5×3.4 ⑩ 0.3×0.25 ⑪ 1.26×2.3 ⑫ 1.5×4.36

⑬ 27×3.4 ⑭ 3.2×1.8 ⑮ 0.4×0.6 ⑯ 7.6×0.5

⑰ 2.87×4.3 ⑱ 0.07×0.8 ⑲ 4.5×0.06 ⑳ 0.2×0.5

Can represent in a math expression using decimal numbers and find the answer.

3 Let's find the area of the rectangle shown on the right.

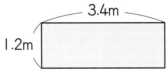
3.4m
1.2m

Understanding the relationship between the multiplier and the product.

4 There is a wire that weighs 4.5 g per meter. Let's find the weight of two wires with a length of 8.6 m and 0.8 m.

Can calculate using the rules of operations.

5 Let's improve the calculation by using the rules of operations. Let's also write the calculation process.

① $7.4 \times 4 \times 2.5$

② $3.8 \times 0.2 \times 5$

③ $4.6 \times 1.9 + 5.4 \times 1.9$

④ $6.8 \times 0.5 - 2.8 \times 0.5$

Supplementary problems
p.150

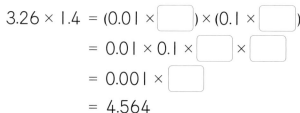

Usefulness and efficiency of learning

1 Let's explain how to calculate 3.26×1.4 by using the calculation of 326×14.

Understanding how to multiply decimal numbers.

$$3.26 \times 1.4 = (0.01 \times \boxed{}) \times (0.1 \times \boxed{})$$
$$= 0.01 \times 0.1 \times \boxed{} \times \boxed{}$$
$$= 0.001 \times \boxed{}$$
$$= 4.564$$

2 Let's find the mistake in the following vertical form and write the correct answer inside the ( ).

Can calculate whole number ×decimal number and decimal number × decimal number in vertical form.

①
```
    4.3
  × 3.14
  ─────
   172
   43
  ─────
  0.602
```
( )

②
```
   0.95
  ×  3.4
  ─────
   380
   285
  ─────
  32.30
```
( )

③
```
    6.4
  ×  2.5
  ─────
   320
   128
  ─────
  1.600
```
( )

3 Instead of multiplying the number by 2.5, a friend added 2.5 to the number and got an answer of 12.3 by mistake. What should have been the answer to the original problem?

Can represent in a math expression using decimal numbers and find the answer.

4 In the following calculations, let's fill in each $\boxed{}$ with an equality or inequality sign.

Understanding the relationship between the multiplier and the product.

① $3.5 \times 3.5 \boxed{} 3.5$ 　　② $3.5 \times 0.1 \boxed{} 3.5$

③ $3.5 \times 0.9 \boxed{} 3.5$ 　　④ $3.5 \times 1 \boxed{} 3.5$

5 Let's improve the calculation. Let's write the calculation process.

Can calculate using the rules of operations.

① $0.5 \times 5.2 \times 8$ 　　② 2.8×15

Which one is a better deal?

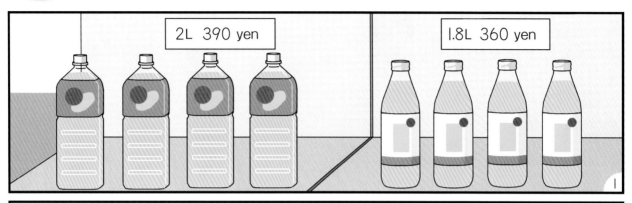

2L 390 yen

1.8L 360 yen

The one with a lot of juice is more expensive.

Which way of buying is a better deal?

Should I find the measure per unit quantity?

The price per liter of juice is to divide 390 yen by 2.

Well, should we divide 360 yen by 1.8?

 Problem Can you solve a calculation when the divisor is a decimal number?

79

7 Division of Decimal Numbers

Let's think about how to calculate and use the rules of operations.

2 L is 390 yen, so I L is...
Nanami

1.8 L is 360 yen, so I L is...
Hiroto

1 Operation of whole number ÷ decimal number

Want to solve Calculation when the divisor is a decimal number

1

How much is the price per liter of juice when 1.8 L is 360 yen?

① Let's use a diagram and table, and think about how to write the math expression to find the price.

Measure per unit quantity — Total measurement

```
              Measure per unit
                 quantity        Total measurement
        0           □              360      (yen)
  Price ├───────────┼───────────────┼───┼───┤
Amount  ├───────────┼───────────────┼───┼───┤
of juice 0          I              1.8  2   (L)
                                  How many units
```

Measure per unit quantity	Total measurement
□yen	360yen
I L	1.8L

How many units

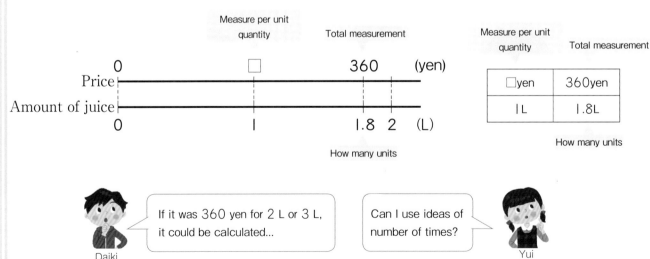

If it was 360 yen for 2 L or 3 L, it could be calculated...
Daiki

Can I use ideas of number of times?
Yui

Purpose Can we think about the price per liter the same as division of whole numbers?

② Let's explain the ideas of the following children.

Daiki's idea

	Price	Amount of juice	Price for 1L

When the amount of juice is 2 L   360 ÷ 2 = 180
3 L   360 ÷ 3 = 120
1.8 L   360 ÷ 1.8 = ☐

Way to see and think

When the "how many units" is a decimal number, can you do the same as with whole numbers?

Yui's idea

If the amount of juice is 1.8 times, the price will be 1.8 times as well.

Way to see and think

Use the idea of number of times.

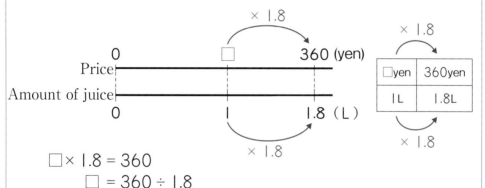

☐ × 1.8 = 360
☐ = 360 ÷ 1.8

Summary

Even if the "how many units" is a decimal number, such as the amount of juice, the calculation of the measurement per unit will be a division in the same way as in whole numbers.

③ About how many yen is the price?

④ Let's think about how to calculate 360 ÷ 1.8.

Purpose What should we do to calculate whole number ÷ decimal number?

⑤　Let's compare the calculation methods of the following children.

Hiroto's idea

Since 1.8 L is the same as 18 sets of 0.1 L,
the cost of 0.1 L of juice is 360 ÷ 18 = 20 (yen).
Since 10 times 0.1 L is the cost of one liter,
the cost of 1 L of juice is 20 × ▢ = ▢ (yen).

```
                              ─── ÷18 ───
                    ┌ 360 ÷ 18
            0       │      ── ×10 ──→    ▢              360 (yen)
    Price  ├────────┼──────────────────┼──────────────┼──────
Amount of juice ├───┼──────────────────┼──────────────┼──────
            0       │      ── ×10 ──→   1              1.8 (L)
                   └ 0.1
                              ─── ÷18 ───
```

Nanami's idea

If I buy an amount of juice that is 10 times 1.8 L,
the price will also become 10 times.
Therefore, I can use the rules of division.

Cost of 1 L when I buy 1.8 L of juice　　360 ÷ 1.8 = ▢ (yen)
　　　　　　　　　　　　　　　　10 times↓　　　↓10 times
Cost of 1 L when I buy 18 L of juice　　3600 ÷ 18 = 200 (yen)

💡 Summary

As for the calculation of whole number ÷ decimal number, the answer
can be found by changing the decimal numbers to whole numbers.

1　The cost of 1.5 L of juice is 240 yen. How many yen is the price of 1 L?
Let's find it out by using the ideas of Hiroto and Nanami shown above.

2

Let's think about how to calculate 360 ÷ 1.8 in vertical form.

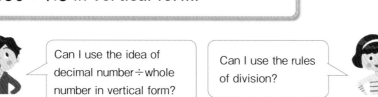

Daiki

Can I use the idea of decimal number ÷ whole number in vertical form?

Can I use the rules of division?

Nanami

Purpose What should we do to calculate whole number ÷ decimal number in vertical form?

① Let's explain how to solve the following calculation in vertical form.

$$1.8 \overline{)3600} \rightarrow 18 \overline{)3600}$$

10 times   10 times

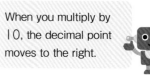

When you multiply by 10, the decimal point moves to the right.

Summary

In division, the quotient does not change if the dividend and divisor are multiplied by the same number. When we divide a number by a decimal number, we can calculate it by changing the dividend and divisor to whole numbers by using the rules of division.

2

A rectangular flowerbed has a length of 2.4 m and an area of 12 m². How many meters is the width?

① Let's write a math expression.

② Let's think about how to calculate in vertical form.

2.4 m   12 m²

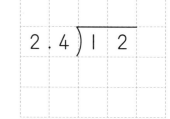

$$2.4 \overline{)1\ 2}$$

3

Let's solve the following calculation in vertical form.

① 9 ÷ 1.8   ② 91 ÷ 2.6   ③ 72 ÷ 4.8

Want to solve

1

5.76 dL of paint was used to paint a wall that is 3.2 m².

How many deciliters of paint are needed to paint a wall that is 1m²?

① Let's think, using a diagram and table, a math expression to find the amount of paint.

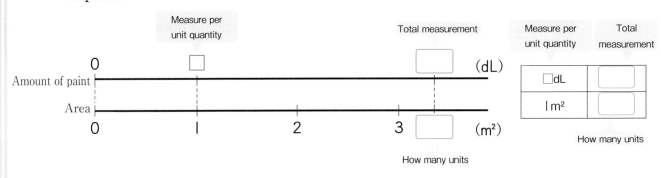

② Let's write a math expression.

Want to explain

③ Let's explain the calculation methods of the following children.

 Daiki's idea

The paint needed for 0.1 m² is

$5.76 ÷ 32 = 0.18$ (dL).

Therefore, the paint needed for 1m² is 10 times that amount,

$0.18 × 10 = $ ☐ (dL)

 Nanami's idea

I will apply the rules of division to change the divisor to a whole number.

$5.76 ÷ 3.2 = $ ☐

10 times ↓   ↓ 10 times

$57.6 ÷ 32 = $ ☐

Can I use the calculation method for whole number ÷ decimal number in vertical form?

Yui

Can I also calculate decimal number ÷ decimal number in vertical form?

Hiroto

⊙ Purpose   Is the calculation of decimal number ÷ decimal number in vertical form the same as before?

④ Let's explain how to calculate $5.76 \div 3.2$ in vertical form.

$$3.2\overline{)5.76}$$

$$3\underset{\text{10 times}}{,}2\overline{)5\underset{\text{10 times}}{,}7.6}$$

$$\begin{array}{r} 1.8 \\ 3,2\overline{)5,7.6} \\ 3\ 2 \\ \hline 2\ 5\ 6 \\ 2\ 5\ 6 \\ \hline 0 \end{array}$$

💡 Summary   Division algorithm of decimal numbers in vertical form

(1) Multiply the divisor by 10, 100, or more to convert it into a whole number. Move the decimal point to the right accordingly.

(2) Multiply also the dividend by the same number as the divisor, and move the decimal point to the right accordingly.

(3) The decimal point of the quotient is aligned at the same place as where the decimal point of the dividend has been moved to.

(4) Then, calculate it as in division of whole numbers.

$$\begin{array}{r} 1.8 \\ 3,2\overline{)5,7.6} \\ 3\ 2 \\ \hline 2\ 5\ 6 \\ 2\ 5\ 6 \\ \hline 0 \end{array}$$

Want to confirm

1 ▶ A rectangular flowerbed has an area of 8.4 m² and a width of 2.8 m. How many meters is the length?

① Let's write a math expression.

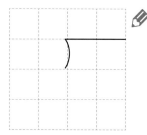

② Let's calculate in vertical form and find the length.

Want to try

2 ▶ Let's solve the following calculations in vertical form.

① $9.52 \div 3.4$ ② $9.88 \div 2.6$ ③ $7.05 \div 1.5$

④ $8.5 \div 1.7$ ⑤ $7.6 \div 1.9$ ⑥ $9.2 \div 2.3$

2 There is a 1.2 m silver wire that weighs 19.2 g and a 0.8 m red wire that weighs 19.2 g. Let's find, for each type, the weight of a wire that is 1 m long.

Purpose What is the size of the quotient when the divisor is a decimal number larger than 1 and smaller than 1?

① Let's find the weight per meter of each wire by writing a math expression.

Silver wire [ ]   Red wire [ ]

② Let's explain about the size of the quotient and size of the dividend using the following number lines.

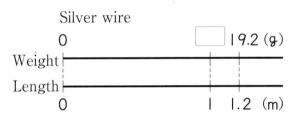

③ Let's calculate 19.2 ÷ [ ], by placing various numbers into [ ], except for 1.2 or 0.8, and let's compare the size with 19.2 as shown on the right.

$19.2 ÷ 0.8 = 24 > 19.2$

$19.2 ÷ 0.4 = $ [ ] [ ] 19.2

$19.2 ÷ 1 = $ [ ] [ ] 19.2

$19.2 ÷ 2 = $ [ ] [ ] 19.2

Summary

When a number is divided by a number larger than 1, the quotient becomes smaller than the dividend.

When a number is divided by a number smaller than 1, the quotient becomes larger than the dividend.

When a number is divided by 1, the quotient becomes the same as the dividend.

Want to confirm

3 Does the quotient become larger than the dividend? Let's calculate and confirm.

① $49 ÷ 0.7$ ② $1.5 ÷ 0.3$ ③ $0.4 ÷ 0.2$

3 There is a 1.5 m metal bar that weighs 4.8 kg. How many kilograms is the weight of a bar that is 1 m long?

① Let's think about it by using diagrams and tables.

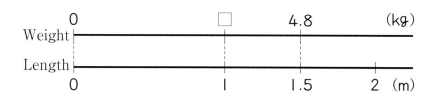

□kg	4.8kg
1m	1.5m

② Let's write a math expression.

③ Daiki thought the calculation in vertical form as shown on the right.

Let's think about how to continue with the division.

```
        3.
  1,5)4 8.
      4 5
        3
```

④ To continue, Daiki thought of 48 as 48.0.

Let's consider the continuation of the calculation on the right and find the answer.

```
        3.
  1,5)4 8.0
      4 5 ↓
        3 0
```

> You can continue with the division of decimal numbers in vertical form, assuming there is a 0 in the lower place.

4 Let's explain how to calculate $3.23 ÷ 3.8$ in vertical form.

Why is the quotient not written in the ones place?

Hiroto

```
         0.8 5
  3,8)3 2.3
      3 0 4
        1 9 0
        1 9 0
            0
```

5 Let's solve the following calculations in vertical form.

① $5.4 ÷ 1.5$ ② $3 ÷ 0.4$ ③ $36.9 ÷ 1.8$ ④ $3.06 ÷ 4.5$

4 Let's think about how to calculate $7.85 \div 3.14$ in vertical form.

$$3.14 \overline{\smash{)}7.85}$$

$$3.14 \overline{\smash{)}7.85} \quad \rightarrow \quad 3.14 \overline{\smash{)}7.85.} \quad \rightarrow \quad \begin{array}{r} 2.5 \\ 3.14 \overline{\smash{)}7.85.} \\ \underline{6\,28} \\ 1\,5\,7\,0 \\ \underline{1\,5\,7\,0} \\ 0 \end{array}$$

100 times 100 times

 6 Let's solve the following calculations in vertical form.

① $4.64 \div 1.45$

② $2.46 \div 2.05$

$$1.45 \overline{\smash{)}4.64}$$

$$2.05 \overline{\smash{)}2.46}$$

 7 Let's solve the following calculations in vertical form.

① $1.75 \div 1.25$ ② $3.24 \div 1.35$ ③ $0.12 \div 0.48$

5

There is a 2.4 m metal bar that weighs 2.72 kg.

How many kilograms is the weight of a bar that is 1 m long?

① Let's write a math expression. [ ]

② The continuation of the calculation is shown on the right. How should we answer?

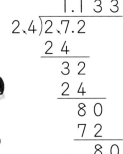

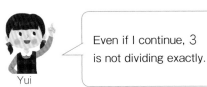
Even if I continue, 3 is not dividing exactly.

Yui

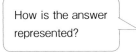
How is the answer represented?

Hiroto

 Purpose How can we represent the quotient when it cannot be divided exactly?

③ As for the quotient, let's find the nearest hundredths place round number by rounding off the thousandths place.

Summary

The quotient can be found by a round number when the division cannot be divided exactly or the number of decimal places increases.

8

There is a 0.3 m wire that weighs 1.6 g. About how many grams is the weight of a wire that is 1 m long?

Let's find the nearest tenths place round number by rounding off the hundredths place.

9

As for the quotient, let's find the nearest hundredths place round number by rounding off the thousandths place.

① 2.8 ÷ 1.7   ② 6.1 ÷ 1.3   ③ 5 ÷ 2.1

④ 61.5 ÷ 8.7   ⑤ 0.58 ÷ 2.3   ⑥ 19.2 ÷ 0.49

6

A tape that is 2.7 m long is cut into pieces that are 0.6 m long. How many 0.6 m long pieces were cut and how many meters of tape remained?

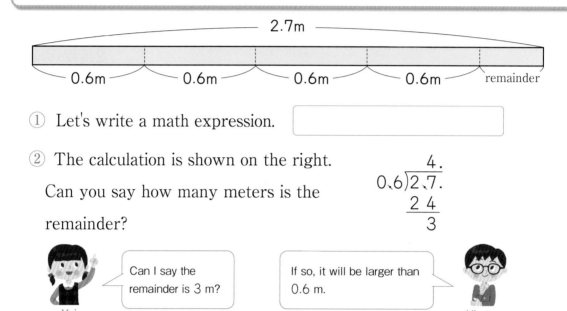

① Let's write a math expression.

② The calculation is shown on the right. Can you say how many meters is the remainder?

$$\begin{array}{r} 4. \\ 0.6\overline{)2.7.} \\ \underline{2\ 4} \\ 3 \end{array}$$

Yui: Can I say the remainder is 3 m?

Hiroto: If so, it will be larger than 0.6 m.

🌱 **Purpose** Where should we place the decimal point of the remainder?

③ Let's explain what you can say about the "three units" in the remainder.

💡 **Summary**

In divisions of decimal numbers in vertical form, the decimal point of the remainder is aligned at the same place as in the original dividend.

$$\begin{array}{r} 4. \\ 0.6\overline{)2.7.} \\ \underline{2|4} \\ 0.3 \end{array}$$

④ Let's confirm the answer.

Dividend = Divisor × Quotient + Remainder

2.7 =   0.6 ×   4 + ☐

10 8 kg of rice will be placed into 1.5 kg bags. How many bags of rice can be filled? How many kilograms of rice will remain?

1

A flowerbed is being watered. Let's think about the area of the flowerbed and the amount of water used.

① 1 m² of the flowerbed is watered with 2.4 L of water. How many liters of water are needed to water an area of 1.5 m²?

It is better to think of a calculation to find the total measurement.

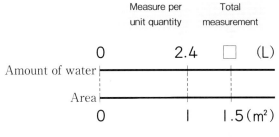

How many units

Math Sentence: 2.4 ⬜ 1.5 = [   ]   Answer: [   ] L

② 4 L of water were used to water an area of 2.5 m². How many liters of water are needed to water an area of 1 m²?

You should find the amount of water per square meter.

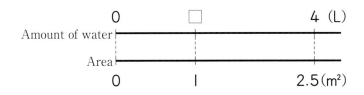

Math Sentence: [   ] ÷ [   ] = [   ]   Answer: [   ] L

③ 2.4 L of water were used to water an area of 1m². How many square meters of the flowerbed can be watered with 8.4 L of water?

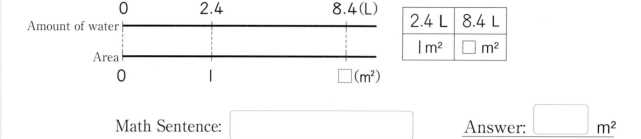

	2.4 L	8.4 L
	1 m²	☐ m²

Math Sentence: [ ]   Answer: [ ] m²

Want to try

1 ▶ Hikaru created the following problem.

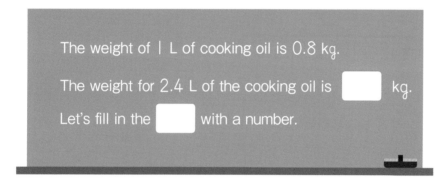

The weight of 1 L of cooking oil is 0.8 kg.

The weight for 2.4 L of the cooking oil is [ ] kg.

Let's fill in the [ ] with a number.

① Let's find the appropriate number for [ ].

② Let's create a multiplication problem by changing the numbers and words.

③ Let's create a division problem by changing the numbers and words.

If it is a problem to find the total measurement, it will be a multiplication.

When calculating the measure per unit quantity, it is a division.

When calculating for a number of units, it will also be a division.

I'd like to make a mixed problem with decimal numbers and whole numbers.

What you can do now

☐ Can calculate whole number ÷ decimal number and decimal number ÷ decimal number in vertical form.

1 Let's solve the following calculations in vertical form. Let's continue to divide with no remainder.

① $80 \div 3.2$　　　　　　② $12 \div 0.6$

③ $39.1 \div 1.7$　　　　　④ $6.5 \div 2.6$

⑤ $29.4 \div 0.3$　　　　　⑥ $4.23 \div 1.8$

⑦ $0.99 \div 1.2$　　　　　⑧ $0.15 \div 0.08$

☐ **Can represent in math expressions using decimal numbers and find the answer.**

2 There is a rectangular flowerbed that has an area of $17.1 m^2$ and a length of 3.8 m. How many meters is the width?

☐ **Understanding the relationship between divisor and quotient.**

3 In the following calculations, let's fill in the ☐ with the equality or inequality signs.

① $125 \div 0.7$ ☐ 125　　　② $125 \div 1.3$ ☐ 125

③ $125 \div 0.89$ ☐ 125　　④ $125 \div 1$ ☐ 125

☐ **Can calculate divisions with remainders.**

4 Let's find the mistake in the following calculations and write the correct answer inside the ( ).

① $4.7 \div 0.6 = 7$ remainder 5　　　② $3.2 \div 5.1 = 6$ remainder 0.14

```
        7
0.6)4.7
      4 2
        5  (        )
```

```
          6
5.1)3.2 0
      3 0 6
        1 4  (        )
```

☐ **Can round off the quotient and find the answer.**

5 Let's answer the following questions when the weight of a 5.2 L kerosene is 3.6 kg. As for the quotient, let's find the nearest hundredths place round number by rounding off the thousandths place.

① How many kilograms is the weight of a 1 L kerosene?

② How many liters is a kerosene that weighs 1kg?

Supplementary problems　p.152

Usefulness and efficiency of learning

1 Let's summarize how to calculate a division of decimal numbers.

$$2.4 \div 0.4 = (2.4 \times \boxed{}) \div (0.4 \times \boxed{})$$
$$= 24 \div \boxed{}$$
$$= \boxed{}$$

☐ Can calculate whole number ÷ decimal number and decimal number ÷ decimal number.

2 Let's write either the × or ÷ sign in the ☐ so that the math sentence is valid.

① 2.8 ☐ 1.4 > 2.8   ② 0.61 ☐ 0.4 < 0.61

③ 7.5 ☐ 0.9 > 7.5   ④ 3.58 ☐ 2.3 < 3.58

☐ Understanding the relationship between divisor and quotient.

3 Let's answer the following questions.

① The weight of a 6.8 m long wire was 1.7 kg.
How many kilograms is the weight of this wire per meter?

② There are 3.4 L of juice. This juice is poured into 0.8 L cups.
How many cups will be filled? How many liters of juice will remain?

☐ Can represent in math expressions using decimal numbers and find the answer.

☐ Can calculate divisions with remainders.

4 0.4 L of paint was used to paint a 5.7 m² wall.

① With 1 L of paint, how many square meters of the wall was painted?

② How many liters of paint are needed to paint a 38 m² wall?
Let's find the nearest tenths place round number by rounding off the hundredths place.

☐ Can round off the quotient and find the answer.

5 There is a 4.8 m wire that weighs 1.2 kg. Let's answer the following questions.

① Let's create a problem that the math expression 4.8 ÷ 1.2 represents and find the answer.

② Let's create a problem that the math expression 1.2 ÷ 4.8 represents and find the answer.

Let's deepen.

I want to make it easier to understand not only by writing but also by drawing diagrams.

Daiki

Deepen.   What's the length of tape?

Want to know

Let's read the sentences and then choose the diagram that represents the relationship.

The length of red tape is 120 cm.

The length of the red tape is 0.6 times the length of white tape.

Want to represent

① Which diagram represents correctly the relationship between the length of the red and the white tapes?　Let's choose from Ⓐ～Ⓓ.

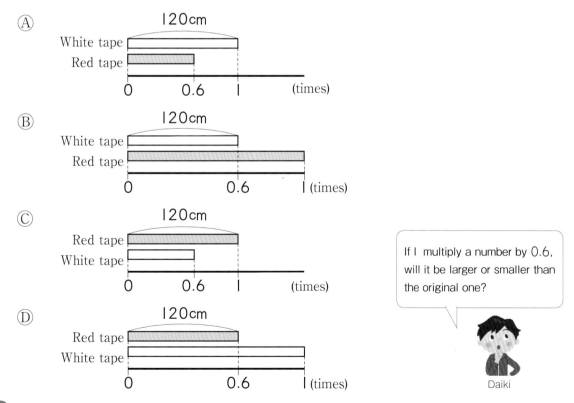

> If I multiply a number by 0.6, will it be larger or smaller than the original one?
>
> Daiki

Want to deepen

Let's write a math expression to find the length of the white tape.

Raising sunflowers.

1 Haruna and friends are raising sunflowers. Let's compare the heights using "times of."

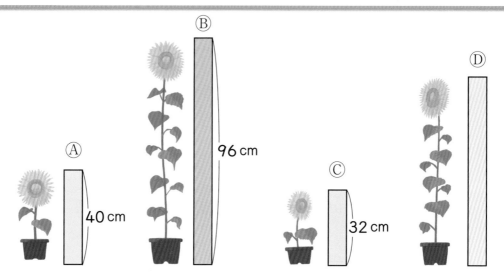

① How many times the height of Ⓐ is the height of Ⓑ?

$$96 \div 40 = \boxed{}$$

Height of Ⓑ   Height of Ⓐ   Times

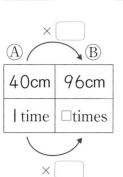

×□
Ⓐ → Ⓑ

40cm	96cm
I time	□times

×□

Way to see and think

If we compare the measure of Ⓑ with Ⓐ, there is a remainder. So, we need to express the answer as a decimal number by dividing the height between 2 and 3 into 10 equal parts.

② How many times of the height of Ⓐ is the height of Ⓒ?

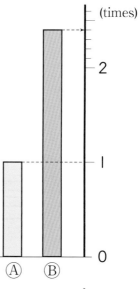

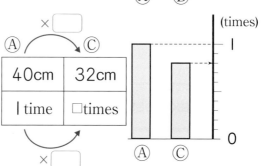

×□
Ⓐ → Ⓒ

40cm	32cm
I time	□times

×□

0.8 times means that when 40 cm is regarded as one, 32 cm represents 0.8 of the reference amount. Times of a decimal number may also be represented by decimal numbers smaller than 1.

Want to try

③ The height of Ⓓ is 2.5 times the height of Ⓒ.

How many centimeters is the height of Ⓓ?

32 × 2.5 = [ ]

Height of Ⓒ   Times   Height of Ⓓ

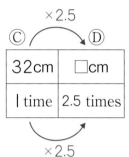

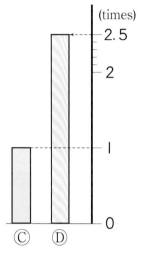

④ How many times the height of Ⓓ is the height of Ⓒ?

32 ÷ [ ] = [ ]

Height of Ⓒ   Height of Ⓓ   Times

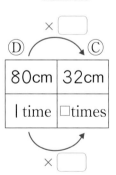

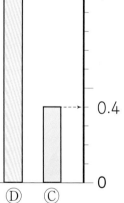

Want to confirm

⑤ Let's represent the height of Ⓑ, Ⓒ, and Ⓓ based on the height of Ⓐ.

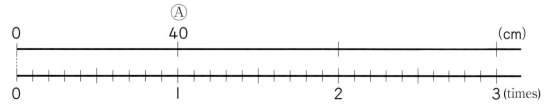

Is the linear motor car fast?

 What should we do to compare the speed?

Measure per Unit Quantity (2)

8 Let's think about how to compare and how to represent which is faster.

1 Speed

Want to explore How to compare

1 The table on the right shows the distance and time from each house to school.

Let's explore who walks fastest.

① Who is faster, Koji or Miyuki?

② Who is faster, Miyuki or Yugo?

Distance and time to school

	Distance (m)	Time (min)
Koji	720	12
Miyuki	660	12
Yugo	660	10

You can compare who is faster if the walking time or walking distance is aligned.

When the time is the same, it is faster to have more distance.

Koji

Miyuki

Walking distance in 1 minute.

When the distance is the same, it is faster to spend less time.

Yugo

Miyuki

Time needed to walk the distance.

③ Who is faster, Koji or Yugo?

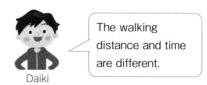

The walking distance and time are different.

Daiki

I can compare if either one is the same ...

Nanami

Purpose What should be done to compare the speed when the walking distance and time are different?

④ Let's compare the ideas of the following children.

Hiroto's idea

I compared the distance they walked per minute.

Koji $720 ÷ 12 =$ ☐ (m)

Yugo $660 ÷ 10 =$ ☐ (m)

Therefore, ☐ is faster.

Yui's idea

I compared the time they walked per meter.

Koji $12 ÷ 720 =$ ☐ (minutes)

Yugo $10 ÷ 660 =$ ☐ (minutes)

Therefore, ☐ is faster.

Summary

Who is faster can be compared by the time per unit of distance or the distance per unit of time.

It is easier to understand the comparison by distance per unit of time because more means faster.

Daiki

Speed is represented as distance per unit of time.

The math sentence to find speed is $$\text{speed} = \text{distance} ÷ \text{time}$$.

⑤ Let's find the speed of each child.

2

The Hakutaka bullet train takes 3 hours to travel 450 km between Tokyo and Kanazawa.

The Hikari bullet train takes 2 hours to travel 366 km between Tokyo and Nagoya.

Which bullet train is faster?

① Let's find the distance that Hakutaka travels per hour.

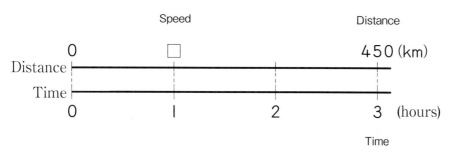

Speed	Distance
□ km	450 km
1 hour	3 hours

Time

② Let's find the distance that Hikari travels per hour.

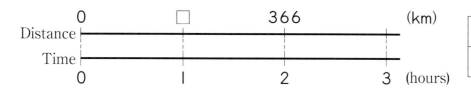

□ km	366 km
1 hour	2 hours

Speed is also a measure per unit quantity. Speed is expressed in various ways depending on the unit of time.

Speed per hour ········Speed expressed by the distance traveled in an hour.

Speed per minute ······Speed expressed by the distance traveled in a minute.

Speed per second······Speed expressed by the distance traveled in a second.

③ How many kilometers per hour is the speed of each bullet train?

Want to try

1   Who is faster, a person that runs 56 m in 8 seconds or a person that runs 60 m in 10 seconds? Let's compare the speed per second.

2   A linear motor car is said to be able to travel 860 km in two hours. Let's find the speed per hour when the linear motor car actually travels at that speed.

3

The news stated that the wind speed of a typhoon is 25 m per second. If a car runs at 54 km per hour, can you say which one is faster?

The speed of wind is also called wind speed.

I can't compare speed per hour with speed per second.

Nanami

Purpose How can we compare the speed per hour, speed per minute, and speed per second?

① Let's explain the ideas of the following children.

Hiroto's idea

I compared the wind speed by changing it to speed per hour.
Since 1 minute is 60 seconds,
$$25 × 60 = 1500$$
the speed becomes 1500 m per minute.
Since 1 hour is 60 minutes,
$$1500 × 60 = 90000$$
the speed becomes 90000 m per hour.
Changing meters to kilometers represents 90 km per hour.

Yui's idea

I compared the speed of the car by changing it to speed per second.
Since 1 hour is 60 minutes,
$$54 ÷ 60 = 0.9$$
the speed becomes 0.9 km per minute, which is equivalent to 900 m per minute.
Since 1 minute is 60 seconds,
$$900 ÷ 60 = 15$$
the speed becomes 15 m per second.

Summary

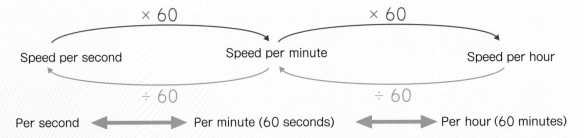

You can compare speed per hour, speed per minute, and speed per second if you align all to one quantity.

Among the following Ⓐ~Ⓒ, which is the fastest?

Ⓐ A car that runs at 30 km per hour.   Ⓑ   A bike that runs at 510 m per minute.

Ⓒ   A 100 m sprinter who runs at 10 m per second.

4 There is a car that runs at 40 km per hour. How many kilometers can it travel in 2 hours? How many kilometers can it travel in 3 hours?

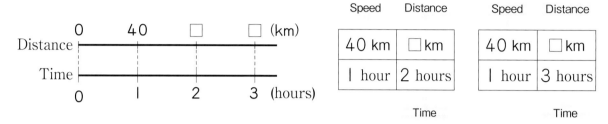

The math sentence to find distance is $\boxed{\textbf{distance = speed × time}}$.

Want to try

4 A cyclist travels at 400 m per minute. How many minutes does it take the cyclist to advance 2400 m?

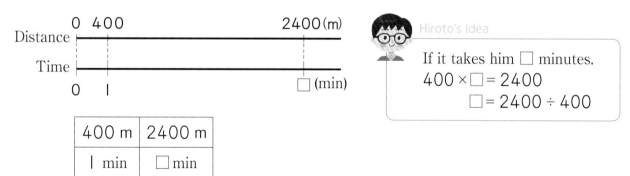

Hiroto's idea

If it takes him ☐ minutes,
$400 × ☐ = 2400$
$☐ = 2400 ÷ 400$

400 m	2400 m
1 min	☐ min

The math sentence to find time is $\boxed{\textbf{time = distance ÷ speed}}$.

Want to confirm

5 Let's answer the following questions about a train that runs at 30 m per second.

① How many meters does it travel in 50 seconds?

② How many minutes will it take to travel 5.4 km?

103

1

There are two machines that pump out 240 L of water in 8 minutes and 300 L of water in 12 minutes respectively. Which machine pumps out more water per minute?

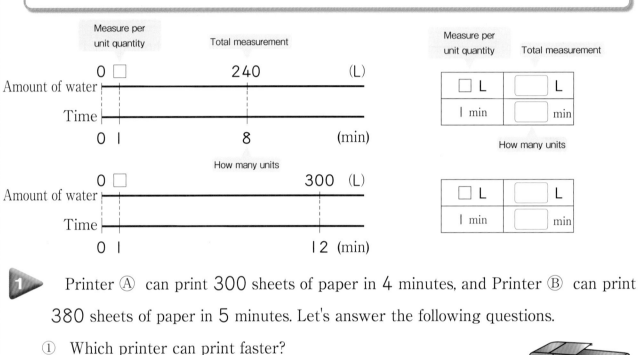

1 ▶ Printer Ⓐ can print 300 sheets of paper in 4 minutes, and Printer Ⓑ can print 380 sheets of paper in 5 minutes. Let's answer the following questions.

① Which printer can print faster?

Ⓐ
☐ sheets	sheets
1 min	min

Ⓑ
☐ sheets	sheets
1 min	min

② How many sheets of paper can be printed in 7 minutes by Printer Ⓐ?

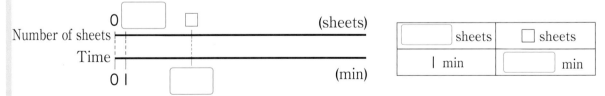

sheets	☐ sheets
1 min	min

③ How many minutes will it take for Printer Ⓑ to print 1140 sheets of paper?

sheets	sheets
1 min	☐ min

What you can do now

☐ **Understanding how to find the speed.**

1 There is a train that travels 210 km in 3 hours and a car that travels 160 km in 2 hours.

  ① How many kilometers per hour is the speed of the train?

  ② How many kilometers per hour is the speed of the car?

☐ **Understanding the relationship between speed per hour, speed per minute, and speed per second.**

2 Let's fill in the following ☐ with numbers.

  ① To change the speed per hour to speed per minute, since 1 hour is ☐ minutes then divide by ☐ .

  ② To change the speed per second to the speed per minute, since 1 minute is ☐ seconds, then multiply by ☐ .

☐ **Understanding how to find the distance.**

3 Let's answer the following questions.

  ① A typhoon is moving at 25 km per hour. How many kilometers will it advance in 12 hours?

  ② It takes 15 minutes to travel by bus from the station to the library. When the bus travels at 42 km per hour, how many kilometers is the distance from the station to the library?

☐ **Understanding how to find the time.**

4 How many seconds will it take for a cheetah to run 180 m if the running speed is 30 m per second?

☐ **Understanding the working speed.**

5 There is a small tractor that cultivates 900 m² in 3 hours. How many square meters will it cultivate in 8 hours?

Supplementary problems ●●●●●●●●➤ p.154

Usefulness and efficiency of learning

1 Let's answer the following questions.

① It took 2 hours and 30 minutes to travel by plane from Naha Airport to Niigata Airport. The air route between the two airports is 1700 km. How many kilometers per hour is the speed of the airplane?

② It takes 12 minutes to walk a distance of 840 m from Reina's house to school. How many meters per minute is Reina's walking speed?

> Understanding how to find the speed.

2 Let's fill in the blanks in the following table and compare the speed.

	Speed per hour	Speed per minute	Speed per second
Racing car		4 km	
Airplane		15 km	
Sound			340 m

> Understanding the relationship between speed per hour, speed per minute, and speed per second.

3 Let's place the following speeds Ⓐ, Ⓑ, and Ⓒ in ascending order.

Ⓐ   A swallow that flies at 5.5 m per second.

Ⓑ   A horse that runs at 1.1 km per minute.

Ⓒ   A car that runs 150 km in 2 hours.

> Understanding the relationship between speed per hour, speed per minute and speed per second.

4 It takes 4 minutes for a car, travelling at 48 km per hour, to go through a tunnel.

① How many meters per minute is the speed of the car?

② What is the length of the tunnel in meters?

③ It took 2 hours and 45 minutes for this car to travel from home to the destination. What is the distance from home to the destination in kilometers?

> Understanding how to find the distance.

5 Takuto's walking speed is 60 m per minute. Let's answer the following questions.

① The distance from Takuto's house to the park is 900 m. How many minutes does it take for Takuto to walk this distance?

② The distance from Takuto's house to his aunt's house is 16.2 km. How many hours and minutes would it take for Takuto to walk this distance?

> Understanding how to find the time.

Let's deepen.

Where can the speed be used in my surroundings?

Daiki

Deepen.

Speed of sound

If you watch fireworks, at first you can watch beautiful fireworks flash and after a while, you can hear the "boom" sound. This sound has its own speed and it takes time for the sound waves to reach you.

Nagaoka City, Niigata Prefecture

Want to know

The speed of sound can be called "sound speed." The speed of sound at a temperature of 0 ℃ is 331 m per second. The table below shows how the sound speed changes depending on the temperature.

Changing in sound speed depending on temperature

Temperature (℃)	0	5	10	15	20	25	30	35	
Speed per second (m)	331	334	337	340	343				

Want to represent

① How many meters per second does the sound speed increase when the temperature increases by 5℃?

② When Daiki and Yui were watching into the sky by a window of the classroom, a lightning flashed and after 6 seconds they heard the sound of the lightning struck. At that moment, the temperature was 15 ℃.

How many meters away did the lightning strike?

③ Yui also saw lightning while she was traveling. She heard the lightning struck 5 seconds after the lightning flashed. The lightning struck 1745 m away from where she was.

What was the temperature in ℃ at that time?

Let's think of an efficient way to distribute supplies.

Want to explore *Which one should we choose?*

Let's think of an efficient way to distribute supplies, using a drone, to the three houses (ⓐ, ⓑ, ⓒ) shown on the map below.

Drone A manual

It can travel **300 m** in **4 minutes**.

The maximum operating time is **15 minutes**.

It can transport **3 kg** of goods at a time.

It costs **3000 yen** for every **15 minutes**.

Drone B manual

It can travel **380 m** in **5 minutes**.

The maximum operating time is **20 minutes**.

It can transport **2 kg** of goods at a time.

It costs **2000 yen** for every **20 minutes**.

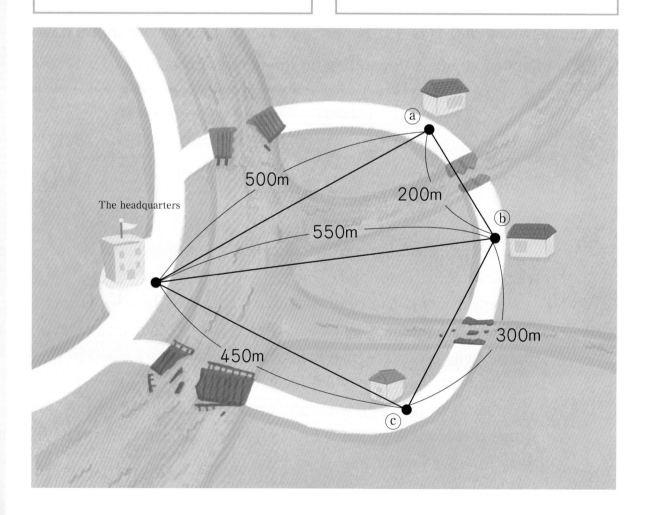

1   I want to distribute the following supplies as soon as possible. How should I use Drone A and Drone B? Let's summarize the plan.

Supplies	Plastic bottles	Cup noodles	Shirts
Address	ⓐ	ⓑ	ⓒ
Quantity	10 bottles (One bottle is 500 g)	10 cups (One cup is 120 g)	2 shirts (One shirt is 320 g)

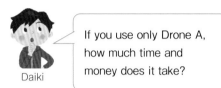

Daiki

If you use only Drone A, how much time and money does it take?

Let's think about drones not only going but also coming back.

2   Let's explain your plan to the group.

For the classmates in each group, let's make an explanatory poster to show how you can use the drone efficiently to distribute the supplies.

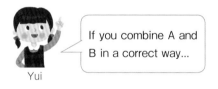

Yui

If you combine A and B in a correct way...

3   Let's present your poster to each other.

In addition, let's compare it with the posters made by other groups and look for examples that you want to imitate or feel good about .

Let's improve and remake your own poster to incorporate the good things of other groups.

01501

Utilizing rule of three on a 4-cell table

1   An area of 2.4m² can be watered with 1 L of water. How many square meters can be watered with 1.5 L of water?

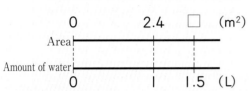

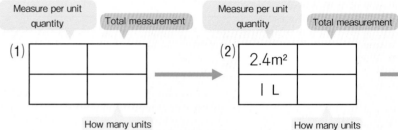

How to create three on a 4-cell table

(1) Write a table with 4 entries.

(2) Since 2.4 m² are watered with 1 L of water, write "1 L" and "2.4 m²" in the left column.

(3) Since you do not know how much area is watered with 1.5 L, write "1.5 L" and "□ m²" in the right column.

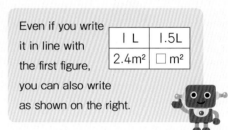

Even if you write it in line with the first figure, you can also write as shown on the right.

1 L	1.5L
2.4m²	□ m²

2   An area of 0.4 m² can be watered with 0.5 L of water. How many square meters can be watered with 1 L of water?

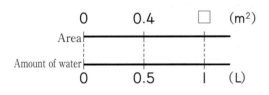

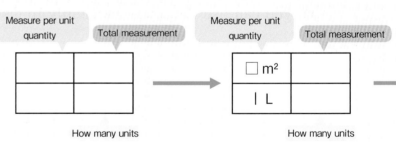

3   An area of 2.4 m² can be watered with 1 L of water. How many liters of water are needed to water an area of 7.2 m²?

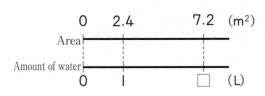

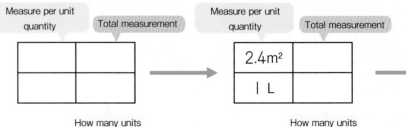

If you think with a word formula

If you think using "times "...

1

Measure per unit quantity	Total measurement
2.4m²	□ m²
1 L	1.5 L

(3)

How many units

$\boxed{\text{Measure per unit quantity}} \times \boxed{\text{How many units}} = \boxed{\text{Total measurement}}$

So,

$$2.4 \times 1.5 = 3.6$$

Answer: 3.6 m²

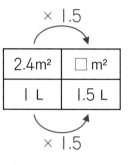

2.4m²	□ m²
1 L	1.5 L

$$2.4 \times 1.5 = 3.6$$

Answer: 3.6 m²

The same unit is placed in the horizontal row.

2

Measure per unit quantity	Total measurement
□ m²	0.4m²
1 L	0.5 L

How many units

$\boxed{\text{Total measurement}} \div \boxed{\text{How many units}} = \boxed{\text{Measure per unit quantity}}$

So,

$$0.4 \div 0.5 = 0.8$$

Answer: 0.8 m²

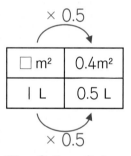

□ m²	0.4m²
1 L	0.5 L

$$\square \times 0.5 = 0.4$$
$$0.4 \div 0.5 = 0.8$$

Answer: 0.8m²

3

Measure per unit quantity	Total measurement
2.4m²	7.2m²
1 L	□ L

How many units

$\boxed{\text{Total measurement}} \div \boxed{\text{Measure per unit quantity}} = \boxed{\text{How many units}}$

So,

$$7.2 \div 2.4 = 3$$

Answer: 3 L

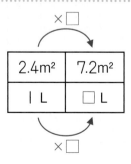

2.4m²	7.2m²
1 L	□ L

$$2.4 \times \square = 7.2$$
$$7.2 \div 2.4 = 3$$

Answer: 3 L

What are the sizes of angles of triangle rulers?

We learned each of the sizes of angles in a triangle ruler.

The size of each angle is 45°, 45°, and 90°.

The sizes for these angles are 30°, 60°, and 90°.

Both have right angles.

1

Ah? If you add all the sizes of the angles, on either triangles, it becomes 180°.

45°

45°   90°

60°

30°   90°

2

Is it the same on other triangles?

3

 Problem   As for the sum of the angles of a triangle, what kind of relationship is there?

112

Angles of Figures

Let's explore about the angles of triangles and quadrilaterals.

1 Sum of the angles of a triangle

Want to explore

1 For the right triangle shown on the right, angle A becomes smaller from 60°, 50°, 40°, ..., and so on, so vertex B gets closer to vertex C. Let's examine about the size of the angles at this time.

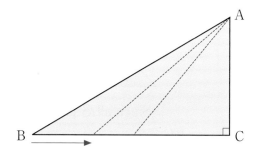

① How does the size of angle B change when the size of angle A decreases by 10°? Let's measure the size of angle B with a protractor and summarize it in the table below.

② Let's find the sum of the size of angle A and angle B.

Angle A (degrees)	60	50	40			
Angle B (degrees)						
Sum (degrees)						

As angle A gets smaller, angle B gets larger.

Daiki

Even if the size of angle A changes, there are things that do not change.

Yui

Purpose Is there a rule for the sum of the three angles of a triangle?

Want to discuss

③ Let's discuss what happens to the sum of the three angles of a triangle.

It looks like the sum of the three angles of a triangle becomes 180°.

But you can't understand unless you examine other triangles.

1 Let's explain the ideas of the following children.

 Hiroto's idea

I cut the three angles, and gathered each vertex into one.

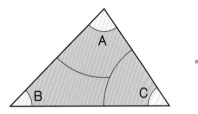 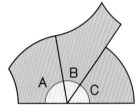

The size of a straight line angle is 180°.

Since all 3 gathered angles became a straight line, their sum is ⬚°.

 Nanami's idea

I placed congruent triangles together, without any gap, to form a continuos pattern.

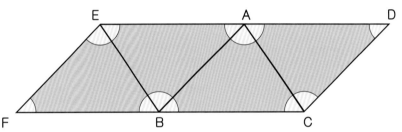

Since all 3 gathered angles at points A or B became a straight line, their sum is ⬚°.

 Yui's idea

I folded the triangle, and attached together the 3 angles.

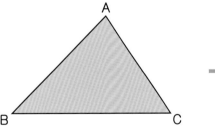 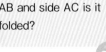

At which point of side AB and side AC is it folded?

Since all 3 gathered angles became a straight line, their sum is ⬚°.

Way to see and think

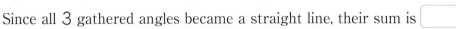

For any triangle, the sum of the three angles is 180°.

The rule can be confirmed in various ways.

2 **Let's find the size of the following angles Ⓐ~Ⓕ by calculations.**

①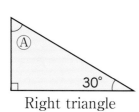

Right triangle

②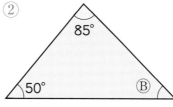

③

Isosceles triangle

④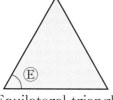

Equilateral triangle

⑤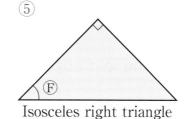

Isosceles right triangle

Want to deepen

2 Let's think about the triangle shown on the right.

① Let's find the sum of angles Ⓐ and Ⓑ.

② Let's find the size of angle Ⓒ.

③ What do you understand about the size of angles Ⓐ, Ⓑ, and Ⓒ?

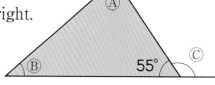

Since Ⓐ+Ⓑ+55°＝180°, then ...

Daiki

Want to confirm

3 Let's find the size of the following angles Ⓐ~Ⓒ by calculations.

①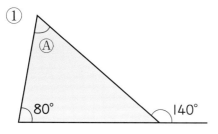

②

③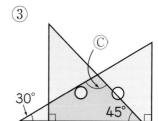

Activity

1

Let's explore about the sum of the four angles of a quadrilateral.

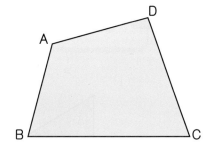

Should I try to measure it the same as a triangle?

Daiki

The same as in triangles, can I understand it by gathering vertices?

Nanami

Purpose As for the sum of the four angles of a quadrilateral, what kind of rule is there?

Way to see and think

① How many degrees is the sum of the four angles of a quadrilateral? Let's explore in various ways.

Can it be examined in the same way as when examining the sum of the angles of a triangle?

Daiki's idea

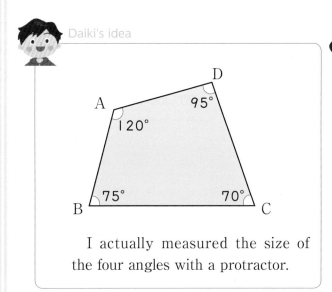

I actually measured the size of the four angles with a protractor.

Nanami's idea

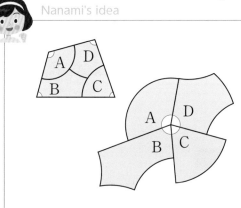

I cut the four angles, and gathered each vertex into one.

② Let's discuss what other methods are available.

③ Let's explain the ideas of the following children.

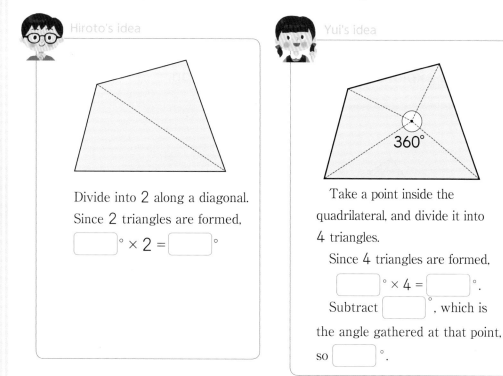

Hiroto's idea

Divide into 2 along a diagonal. Since 2 triangles are formed,

[   ]° × 2 = [   ]°

Yui's idea

Take a point inside the quadrilateral, and divide it into 4 triangles.

Since 4 triangles are formed,

[   ]° × 4 = [   ]°.

Subtract [   ]°, which is the angle gathered at that point,

so [   ]°.

Way to see and think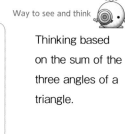

Thinking based on the sum of the three angles of a triangle.

Summary

For any quadrilateral, the sum of the four angles is 360°.

---

1 Let's confirm, through the tessellation of congruent quadrilaterals, that the sum of the four angles of a quadrilateral is 360°.

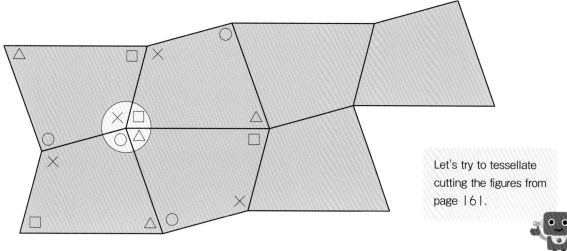

Let's try to tessellate cutting the figures from page 161.

2 I would like to verify that the sum of the four angles of the figure shown on the right is also 360°. Cut the figures from page 161. Let's explore by creating a tessellation.

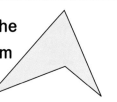

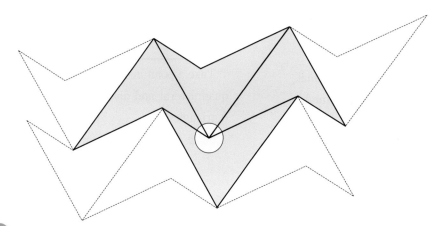

Way to see and think

What kind of angle is gathered at one point?

2 The verification, that the sum of the four angles of the figure in **2** is 360°, was done as follows. Let's explain the ideas of the following children.

Daiki's idea

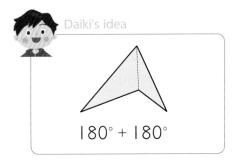

180° + 180°

Nanami's idea

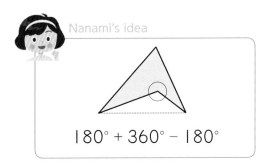

180° + 360° − 180°

 Let's find the size of the following angles Ⓐ〜Ⓒ by calculations.

①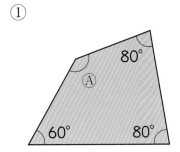

②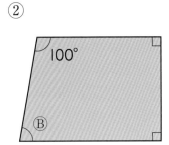

③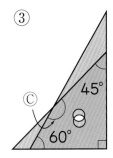

Want to think : Sum of the angles of a pentagon

Activity

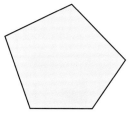

1

> A figure enclosed by five straight lines is called a pentagon. Let's think about how to find the sum of the five angles of a pentagon.

Daiki: Can I think of it as the tessellations we did before?

The same as with quadrilaterals, if I divide into triangles...

Nanami

Purpose How many degrees is the sum of the five angles of a pentagon?

Want to explain

① Let's explain the ideas of the following children.

Way to see and think

Thinking based on the sum of the angles of triangles or quadrilaterals.

 Hiroto's idea

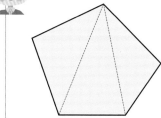

If you draw straight lines from one vertex to a vertex that is not adjacent, it can be divided into ☐ triangles.
Therefore, $180° \times$ ☐ $=$ ☐ $°$

 Yui's idea

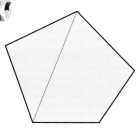

If you draw one straight line from one vertex to a vertex that is not adjacent, it can be divided into a triangle and quadrilateral.
Therefore, $180° +$ ☐ $° =$ ☐ $°$

 Nanami's idea

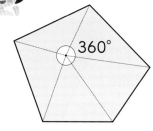
360°

Take a point inside the pentagon and divide it into 5 triangles. Since 5 triangles are formed, ☐ $° \times 5 =$ ☐ $°$.
Subtract ☐ $°$, which is the angle gathered at that point, so ☐ $°$.

Summary

For any pentagon, the sum of the five angles is 540°.

Want to explain

 1 Daiki tried to tessellate the pentagon the same way as in the case of the quadrilateral but it was not successful. Let's explain the reason why it was not successful.

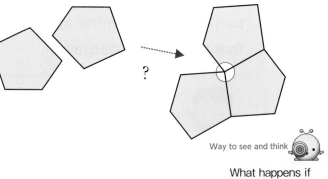

Way to see and think

What happens if you gather 5 angles at one vertex?

In order to tessellate, the sum of the angles gathered at one vertex must be 360°.

Want to think Sum of the angles of a hexagon

2 A figure enclosed by six straight lines is called a hexagon. Let's think about how to find the sum of the six angles of a hexagon.

It seems like you can think of it as before.

Yui

Can I find any rule?

Hiroto

Purpose How many degrees is the sum of the six angles of a hexagon?

① Let's explain how to find the sum of the six angles of a hexagon using the hexagon shown on the right.

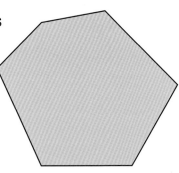

Summary

For any hexagon, the sum of the six angles is 720°.

A figure that is enclosed only by straight lines, such as triangles, quadrilaterals, pentagons, hexagons, etc., is called a **polygon**. In a polygon, each straight line that connects any two vertices that are not adjacent is called a **diagonal.**

Want to extend

 2 For various polygons, let's summarize the sum of the angles using the number of triangles that can be separated by diagonals drawn from one vertex.

	Triangle	Quadrilateral	Pentagon	Hexagon	Heptagon	Octagon	Nonagon
Number of triangles	(1)	2	3	4			
Sum of the angles	180°	360°	540°	720°			

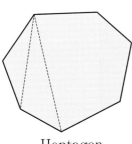

Way to see and think

How does the sum of the angles change as the number of triangles increases?

Heptagon   Octagon   Nonagon

180° × ☐ = ☐°   180° × ☐ = ☐°   180° × ☐ = ☐°

Want to confirm

 3 The following math expressions represent how to find the sum of the five angles of a pentagon. Choose the figure from Ⓐ～Ⓓ that matches with the math expressions from ①～④.

① 180° + 360°   ② 180° × 3   ③ 180° × 4 − 180°   ④ 180° × 5 − 360°

Ⓐ   Ⓑ   Ⓒ   Ⓓ

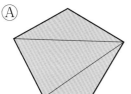
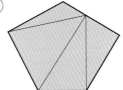
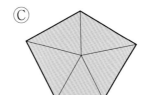
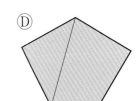

What you can do now

☐ **Understanding the sum of the angles of a triangle.**

1 Let's find the size of the angles Ⓐ and Ⓑ by calculations.

①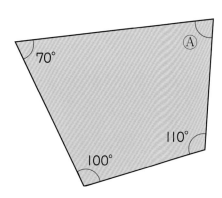

②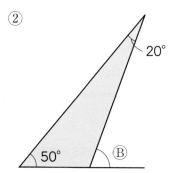

☐ **Understanding the sum of the angles of a quadrilateral.**

2 Let's find the size of the angles Ⓐ and Ⓑ by calculations.

①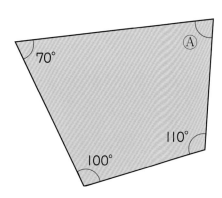

②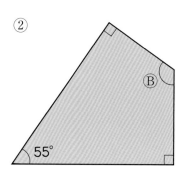

☐ **Undestanding how to find the sum of the angles of a polygon.**

3 Let's fill in each ☐ with a number.

Find the sum of the angles of a heptagon.

If you draw diagonals from one vertex, it can

be divided into ☐ triangles.

Therefore, 180° × ☐ = ☐ °

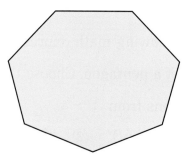

Heptagon

Supplementary problems
p.155

Usefulness and efficiency of learning

1 Let's find the size of the following angles Ⓐ~Ⓕ by calculations.

Understanding the sum of the angles of a triangle.

Understanding the sum of the angles of a quadrilateral.

Understanding how to find the sum of the angles of a polygon.

①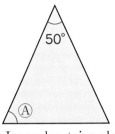

50°

Ⓐ

Isosceles triangle

②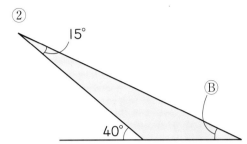

15°

Ⓑ

40°

③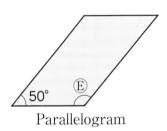

130°

60°

Ⓒ

④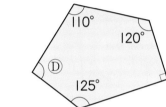

110°

120°

Ⓓ

125°

⑤

50°

Ⓔ

Parallelogram

⑥ Hexagon created with 6 equilateral triangles

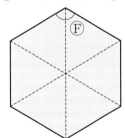

Ⓕ

2 Let's find the size of the angles Ⓐ~Ⓓ by calculations. These were made by overlapping triangle rulers.

Can use the sum of the angles of a polygon.

①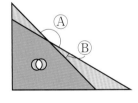

Ⓐ

Ⓑ

②

Ⓓ

Ⓒ

3 Let's explain how to find the sum of the angles of an octagon using the following words.

> vertex, diagonal, triangle, 180°

Understanding how to find the sum of the angles of a polygon.

Let's divide into teams.

Divide into two teams, red and white, to play dodgeball.

There are 32 students in the class.

In what way should we divide?

Since we just need to divide in half, $32 \div 2 = 16$, each team should have 16 students.

Then, I wonder if they should be divided into $1 \sim 16$ and $17 \sim 32$ by attendance numbers.

There is also another way to divide. First person is red, second person is white, third person is red,..., and so on. You can divide them in order.

Problem When categorizing numbers in order, is there any rule?

10 Multiples and Divisors

Let's think about the multiples and divisors.

1 Even numbers and odd numbers

Want to explore   How to categorize whole numbers

1   Divide 32 children in the class into red and white teams. When you divide "red, white, red, white, ... " by attendance number order, what kind of number gathering can you identify for each team?

① In what kind of numbers were the red and white teams gathered?

Let's write the attendance number and complete the following table.

red	1	3	5			
white	2	4	6			

Daiki: Red and white numbers are increasing by 2.

What kind of arrangement is it?

Nanami

Purpose Is there any rule for each number consistency?

② Let's try to divide the attendance number by 2 for the red and white teams.

Way to see and think
There are times with 1 as remainder and times with no remainder.

Red team
$1 \div 2 = 0$ remainder 1
$3 \div 2 = 1$ remainder 1
$5 \div 2 = 2$ remainder 1
⋮

White team
$2 \div 2 = 1$
$4 \div 2 = 2$
$6 \div 2 = 3$
⋮

③ Which team will the children with attendance number 11 and 18 join?

④ Let's think what kind of categorizing rules there are for each team.

For whole numbers, the numbers that are divisible by 2 are called **even number**s, and numbers that are not divisible by 2 are called **odd number**s. 0 is an even number.

Also, even and odd numbers have the following representation.

Even number: number multiplied by 2.   Odd number: Add 1 to a number multiplied by 2.

$$2 \times \square$$ $$2 \times \square + 1$$

Want to find

1 How are even and odd numbers aligned? Let's explore and write a ○ on even numbers and □ on odd numbers in the following number line. Are there any whole numbers that do not enter as even or odd numbers?

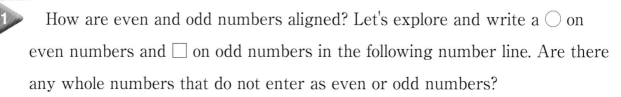

Summary

Whatever whole number can be categorized as either even or odd number.

Whole numbers	
Even numbers	Odd numbers
0, 2, 4, 6, …	1, 3, 5, 7, …

Want to confirm

2 Let's categorize the following whole numbers as even or odd numbers.

0   7   28   29   30   98   99   100   217   218   1234

Want to find in our life

3 From your surroundings, let's find occasions where even and odd numbers are used.

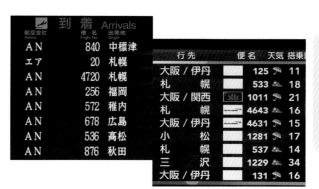

The number of flights to Tokyo is an even number, and the number of flights from Tokyo is an odd number.

126

【Rules of the Clap Number Game】

Make a circle with everyone. Decide the person that will start. Say out loud the numbers in order starting from 1, and clap your hands on each set of the "Clap Number." For example, when the "Clap Number" is 3, at every set of 3 people, clap the hands while saying out loud the corresponding number.

2 Multiples and common multiples

Want to think   Multiples

Group A decided the "Clap Number" is 3. Let's think about which numbers will have clapping hands.

① Let's write the numbers in the table shown on the right, and color the number that will have clapping hands.

② As for the colored numbers, let's discuss which numbers are gathered.

1	2	3	4	5	6	7	8	9	10
11	12	13	14	15	16	17	18	19	20
21	22								

The colored numbers became 3, 6, ...

It became the numbers that come out in the row of 3 in the multiplication table.

A number resulting from the multiplication of a whole number by 3, such as 3×1, 3×2, 3×3, ..., is called a **multiple** of 3.
   0 of 0×3 is not a multiple of 3.

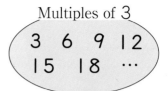

Multiples of 3

3   6   9   12
  15   18   ···

③ Let's place a ◯ on the multiples of 3 in the number line ⓑ at the bottom of this page.

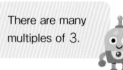

There are many multiples of 3.

Want to confirm

1 Let's place a ◯ on the multiples of 2 and 4 in the respective number lines ⓐ and ⓒ at the bottom of this page.

2 Let's answer the following questions when stacking boxes with a height of 5 cm.

Cookies

5cm

① How many cm is the total height when 6 boxes are stacked?

② The total height became multiples of what number?

3 Let's find the smallest 5 numbers that are multiples of the following.

① Multiple of 7   ② Multiple of 8   ③ Multiple of 9

Want to try

4 Let's think about multiples of what numbers are the following numbers?

① 10   ② 12   ③ 24

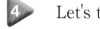

ⓐ Multiples of 2
0 1 2 3 4 5 6 7 8 9 10 11 12 13 14 15 16 17 18 19 20 21 22 23 24 25 26 27

ⓑ Multiples of 3
0 1 2 3 4 5 6 7 8 9 10 11 12 13 14 15 16 17 18 19 20 21 22 23 24 25 26 27

ⓒ Multiples of 4
0 1 2 3 4 5 6 7 8 9 10 11 12 13 14 15 16 17 18 19 20 21 22 23 24 25 26 27

That's it

How are the multiples aligned?

The multiples of **2** were colored in the table with numbers from **1** to **100**.

① How are the multiples of **2** aligned?

② Let's explore about the multiples of **3**.

Multiples of 2

1	2	3	4	5	6	7	8	9	10
11	12	13	14	15	16	17	18	19	20
21	22	23	24	25	26	27	28	29	30
31	32	33	34	35	36	37	38	39	40
41	42	43	44	45	46	47	48	49	50
51	52	53	54	55	56	57	58	59	60
61	62	63	64	65	66	67	68	69	70
71	72	73	74	75	76	77	78	79	80
81	82	83	84	85	86	87	88	89	90
91	92	93	94	95	96	97	98	99	100

Multiples of 3

1	2	3	4	5	6	7	8	9	10
11	12	13	14	15	16	17	18	19	20
21	22	23	24	25	26	27	28	29	30
31	32	33	34	35	36	37	38	39	40
41	42	43	44	45	46	47	48	49	50
51	52	53	54	55	56	57	58	59	60
61	62	63	64	65	66	67	68	69	70
71	72	73	74	75	76	77	78	79	80
81	82	83	84	85	86	87	88	89	90
91	92	93	94	95	96	97	98	99	100

Let's color the multiples of 3.

③ The multiples of what number were colored in the tables with numbers from **1** to **100**?

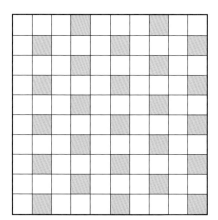

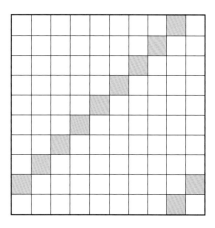

It is beautifully aligned.

Daiki

Let's explore other multiples.

28 29 30 31 32 33 34 35 36 37 38 39 40 41 42 43 44 45 46 47 48 49 50 51 52 53 54 55 56 57 58 59 60

28 29 30 31 32 33 34 35 36 37 38 39 40 41 42 43 44 45 46 47 48 49 50 51 52 53 54 55 56 57 58 59 60

28 29 30 31 32 33 34 35 36 37 38 39 40 41 42 43 44 45 46 47 48 49 50 51 52 53 54 55 56 57 58 59 60

2 In the Clap Number Game, Group A clap hands when it is a multiple of 3, and Group B clap hands when it is a multiple of 4. Let's discuss what you noticed by actually playing the Clap Number Game.

Group A clap hands as follows:

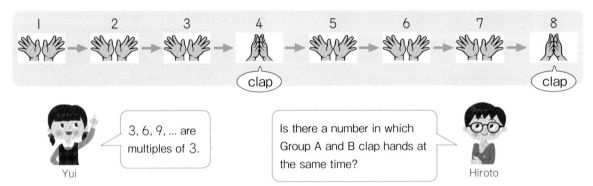

Group B clap hands as follows:

3, 6, 9, ... are multiples of 3.

Yui

Is there a number in which Group A and B clap hands at the same time?

Hiroto

Purpose On which number are they clapping hands at the same time?

Want to explore

① In the following table, let's write ○ for the number in which Group A and B clap hands, and let's write × for the number in which they do not clap hands.

	1	2	3	4	5	6	7	8	9	10	11	12
Group A	×	×	○	×								
Group B	×	×	×	○								

② When do both groups clap hands at the same time?

③ When they continue the game, when is the second time that both groups clap hands at the same time?

A number that is a multiple of both 3 and 4 is called a **common multiple** of 3 and 4.
The smallest of all common multiples is called **the least common multiple**.

You can use the number lines from p.128 and p.129.

Want to think

④ Let's find 5 common multiples of 3 and 4 in ascending order. Also, what is the least common multiple of 3 and 4?

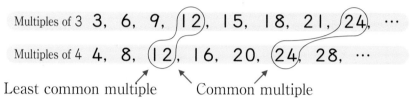

Multiples of 3 3, 6, 9, (12), 15, 18, 21, (24), ...

Multiples of 4 4, 8, (12), 16, 20, (24), 28, ...

Least common multiple   Common multiple

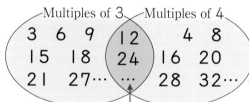

Common multiples

Way to see and think

Similarly, as there are many multiples of a number, there are also many common multiples.

Summary

When you clap hands in multiples of 3 and 4, clapping hands at the same time in 12, 24, ..., etc., are the common multiples of 3 and 4.

That's it

Make tapes that find common multiples.

When a tape with holes at multiples of 2 and a tape with holes at multiples of 3 are stacked, the number of places where they overlap are common multiples of 2 and 3. Let's actually make tapes with holes on various multiples and find common multiples.

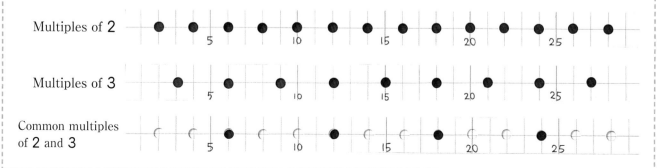

Activity

3

Let's think about how to find the common multiples of 4 and 6.

Just write a multiple of each number.

Daiki

Can I do it more easily?

Nanami

Want to explain

① Let's explain the way of thinking of the following 4 children.

Daiki's idea

Multiples of 4   4, 8, (12), 16, 20, (24), 28, 32, (36), 40, ⋯

Multiples of 6   6, (12), 18, (24), 30, (36), 42, 48, 54, 60, ⋯

Yui's idea

Multiples of 4   4,   8,   12, 16,   20, 24, 28, 32, 36, ⋯

　　　　　　　×   ×   ○   ×   ×   ○   ×   ×   ○

Nanami's idea

Multiples of 6   6, 12, 18, 24, 30, 36, 42, 48, ⋯

　　　　　　　×   ○   ×   ○   ×   ○   ×   ○

Hiroto's idea

4, 8, (12)　　　　12 × 2 = (24)

6, (12)　　　　　12 × 3 = (36)

② Which is the least common multiple of 4 and 6?

③ Using Hiroto's idea, lets find 5 common multiples of 4 and 6 in ascending order.

The least common multiple of 4 and 6 is 12.
The common multiples of 4 and 6 are multiples of the least common multiple.

2 times   3 times   4 times

12　　24　　36　　48⋯

5 Let's find 4 common multiples of the following set of numbers in ascending order.

① (5, 2)　　　② (3, 9)　　　③ (6, 9)

Activity

 6 Nanami considered the least common multiple and common multiples of 2 and 6 as follows. Is this way of thinking correct? In the case it's not, let's write down the reasons.

 Nanami's idea

> The least common multiple of 3 and 4 is 12. This can be found by 3 × 4 = 12. Therefore, since 2 × 6 = 12, the least common multiple of 2 and 6 is also 12. Thus, the common multiples of 2 and 6 are multiples of 12.

Common multiple of 3 numbers

4 **Let's find the common multiples of 2, 3, and 4.**

① Let's place a ○ on the multiples of 2, 3, and 4 in each of the following number lines.

Multiples of 2
0 1 2 3 4 5 6 7 8 9 10 11 12 13 14 15 16 17 18 19 20 21 22 23 24

Multiples of 3
0 1 2 3 4 5 6 7 8 9 10 11 12 13 14 15 16 17 18 19 20 21 22 23 24

Multiples of 4
0 1 2 3 4 5 6 7 8 9 10 11 12 13 14 15 16 17 18 19 20 21 22 23 24

② Which is the least common multiple of 2, 3, and 4?

③ Let's find 3 common multiples of 2, 3, and 4 in ascending order.

 7 Let's find 4 common multiples of the following set of numbers in ascending order. Also, let's find the least common multiple.

① (3, 5, 6)　　　② (4, 7, 14)　　　③ (5, 10, 15)

5

Let's make a square by aligning 5 cm long and 6 cm wide rectangular paper in the same direction as shown on the right.

Let's answer the following questions about the resulting square.

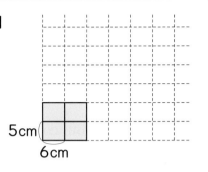

5cm

6cm

① The length and width became multiple of which number?

In order to make a square, the length and width should be equal.

Yui

② How many cm is the length of the side of the smallest square created?

③ Let's find the length of the sides of three squares created in ascending order.

8 Boxes of cookies with a height of 6 cm and boxes of chocolates with a height of 8 cm are piled up. What is the height when the heights of both piles of boxes are equal for the first time?

6cm

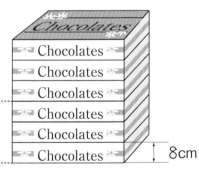

8cm

9 There is a metronome that runs every 12 seconds and another metronome that runs every 20 seconds. If both start at the same time, after how many seconds will the two metronomes concur again?

Want to explore

1
Tessellate squares of the same size so that there are no gaps in a 12 cm long and 18 cm wide rectangle. When you can do this, how many cm is the length of one side of the squares?

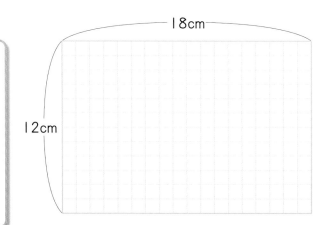

18cm

12cm

A square with a side length of 1 cm can be tessellated.

First, let's find a square that can be tessellated vertically.

Can other squares be tessellated?

Later, let's find a square that can be tessellated horizontally.

Way to see and think

① Just for vertical tessellation, in exactly 12 cm, how many cm is the length of one side of the squares?

Let's think separately the horizontal and vertical tessellation.

In vertical tessellation by 12 cm without gaps, the possible lengths of one side of the squares are: 1 cm, 2 cm, 3 cm, 4 cm, 6 cm, and 12 cm.

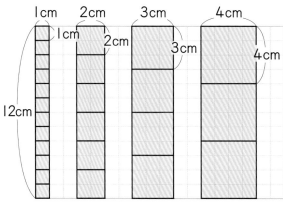

② Let's try to divide 12 by 1, 2, 3, 4, 6, and 12 one by one.

The whole numbers, such as 1, 2, 3, 4, 6, and 12, by which 12 can be divided with no remainder are called **divisor**s of 12.

Divisors of 12
1   2   3
4   6   12

12 is a multiple of 1, 2, 3, 4, 6, and 12.

③ What can you tell when the divisors of 12 are paired as shown on the right?

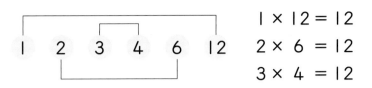

$$1 \times 12 = 12$$
$$2 \times 6 = 12$$
$$3 \times 4 = 12$$

For any whole number, 1 and the number itself are divisors.

④ Just for horizontal tessellation, in exactly 18 cm, how many cm is the length of one side of the squares?

In horizontal tessellation by 18 cm without gaps, the possible lengths of one side of the squares are: 1 cm, 2 cm, 3 cm, 6 cm, 9 cm, and 18 cm.

  1, 2, 3, 6, 9, and 18 are the divisors of 18.

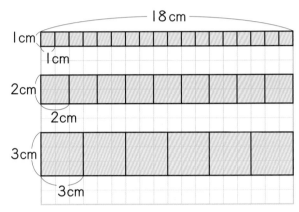

Want to find

⑤ How many cm is the length of one side of a square that can be tessellated vertically and horizontally without gaps?

Vertical tessellation........ 1 2 3 4 6 12 (cm)
Horizontal tessellation... 1 2 3 6 9 18 (cm)

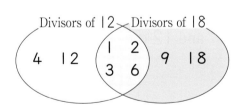

  The numbers that are divisors of both 12 and 18 are called **common divisor**s of 12 and 18. The largest of all common divisors is called **the greatest common divisor**.

Divisors of 12   Divisors of 18
4   12   1   2   9   18
     3   6

⑥ Let's write all the common divisors of 12 and 18. Also, which is the greatest common divisor of 12 and 18?

Want to confirm

1 ▷ Let's find all the divisors of 8 and 36. Also, let's find all the common divisors of 8 and 36.

2

Let's think about how to find the common divisors of 18 and 24.

Just write all the divisors.

Daiki

I wonder if it can be done like when I find the common multiple.

Nanami

Want to explain

① Let's explain the way of thinking of the following 2 children.

Hiroto's idea

Divisors of 18 1, 2, 3, 6, 9, 18
Divisors of 24 1, 2, 3, 4, 6, 8, 12, 24

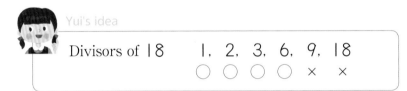

Yui's idea

Divisors of 18   1, 2, 3, 6, 9, 18
                  ○ ○ ○ ○ × ×

In the case of 18 and 24, you find it from the divisors of the smaller number 18.

② Let's find all the common divisors of 18 and 24. Also, which is the greatest common divisor?

③ Let's say what you noticed by looking at the common divisors and the greatest common divisor of 18 and 24.

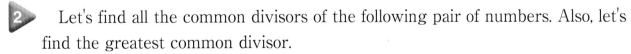

The common divisors of 18 and 24 are divisors of the greatest common divisor of 18 and 24.

Want to confirm

2 Let's find all the common divisors of the following pair of numbers. Also, let's find the greatest common divisor.

① (8, 16)   ② (15, 20)   ③ (12, 42)   ④ (13, 9)

3 I would like to hand out 8 pens and 12 notebooks, such that each child receives the same number of items per type with no remainder. In how many children can I distribute successfully?

3 **Let's find the common divisiors of 6, 9, and 12.**

① Let's place a ◯ on the divisors of 6, 9, and 12 in each of the following number lines.

Divisors of 6
0 1 2 3 4 5 6

Divisors of 9
0 1 2 3 4 5 6 7 8 9

Divisors of 12
0 1 2 3 4 5 6 7 8 9 10 11 12

② Which is the greatest common divisor of 6, 9, and 12?

③ Let's find all the common divisors of 6, 9, and 12.

 Let's find all the common divisors of the following set of numbers. Also, let's find the greatest common divisor.

① (8, 18, 24)   ② (6, 12, 27)   ③ (7, 16, 23)

 Let's explore about the relationship between multiples and divisors.

① Let's align 18 square cards in a rectangular shape and find the divisors of 18.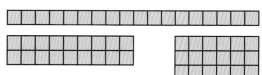

② Is 18 a multiple of the divisors found in ①?

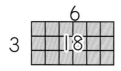

- 3 and 6 are divisors of 18.
- 18 is a multiple of 3 and 6.

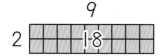

- 2 and ☐ are divisors of 18.
- 18 is a multiple of 9 and ☐.

18 is a multiple of 3, 3 is a divisor of 18.

3 ⟶ multiple ⟶ 18
3 ⟵ divisor ⟵ 18

What you can do now

☐ **Understanding about even and odd numbers.**

1 Let's categorize the following whole numbers as even or odd numbers.

0   1   2   4   8   9   13   17
68   108   493   78542

Even numbers	Odd numbers

☐ **Can find multiples and divisors.**

2 Let's find 3 multiples of the following numbers in ascending order. Also,

let's find all the divisors.

① 6   ② 13   ③ 16   ④ 24

☐ **Can find common multiples and the least common multiple.**

3 Let's find 3 common multiples of the following set of numbers in ascending order.

Also, let's find the least common multiple.

① (3, 6)   ② (5, 7)

③ (6, 10)   ④ (8, 12)

☐ **Can find common divisors and the greatest common divisor.**

4 Let's find all the common divisors of the following set of numbers.

Also, let's find the greatest common divisor.

① (9, 15)   ② (12, 24)

③ (30, 42)   ④ (28, 42)

☐ **Can solve problems using common multiples and common divisors.**

5 At a station, a train departs every 12 minutes and a bus departs every

8 minutes. At 9 a.m., the train and the bus departed at the same time. At

what hour and minutes is the next time that both depart together?

Supplementary problems
•••••••• p.157

Usefulness and efficiency of learning

1 Even and odd numbers can be represented as shown in the figure on the right. Let's use the diagram on the right to explain that the sum of two odd numbers becomes an even number.

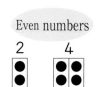

Even numbers

2   4

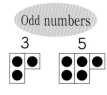

Odd numbers

3   5

☐ Understanding about even and odd numbers.

2 Let's answer the following questions about the numbers from 1 to 50.

1	2	3	4	5	6	7	8	9	10
11	12	13	14	15	16	17	18	19	20
21	22	23	24	25	26	27	28	29	30
31	32	33	34	35	36	37	38	39	40
41	42	43	44	45	46	47	48	49	50

① Let's find all the multiples of 3.

② Let's find all the multiples of 7.

③ Let's find all the common multiples of 3 and 7.

☐ Can find multiples.

☐ Can find common multiples.

3 I would like to divide 24 candies by some people such that there is no remainder and each receives the same number. How many people will receive how many candies if I divide without having remainder? Let's answer all possible ways of dividing.

☐ Can find divisors.

4 Let's make a square by aligning 6 cm long and 9 cm wide rectangle as shown on the right. How many cm is the length of one side of the smallest square? Also, how many rectangles were aligned in total?

9cm

6cm

☐ Can find common divisors and the greatest common divisor.

5 Cut out squares of the same size so that there is no waste from a 12 cm long and 30 cm wide graphing paper. How many cm is the length of one side of the largest square? Also, how many squares with that side length can you cut out?

☐ Can solve problems using common multiples and common divisors.

Let's deepen.

When can I use multiples and divisors?

Daiki

Deepen.

Can you weigh?

I would like to use a balance to weigh various objects. However, when weighing, I can only use two types of weights: 3 g weight and 10 g weight. Also, the weight can be placed only on the right dish. What kind of weight can you weigh?

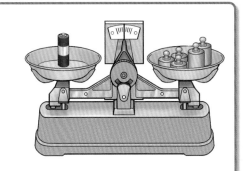

I can weigh multiples of 3 and multiples of 10.

Nanami

I cannot weigh anything that is either.

Daiki

① When a battery was weighed, three 3 g weights and two 10 g weights were balanced. At that time, how many grams is the weight of the battery?

② Which is the weight that cannot be weigh with this balance? Choose from the following Ⓐ～Ⓓ.

    Ⓐ 9 g   Ⓑ 10 g   Ⓒ 11 g   Ⓓ 12 g

③ Is there any other weight that cannot be weighed? Let's investigate using the table on the right by placing a ◯ on the weights you can weigh and placing a × on the weights you cannot weigh.

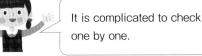

It is complicated to check one by one.

Yui

1	2	3	4	5	6	7	8	9	10
11	12	13	14	15	16	17	18	19	20
21	22	23	24	25	26	27	28	29	30
31	32	33	34	35	36	37	38	39	40
41	42	43	44	45	46	47	48	49	50
51	52	53	54	55	56	57	58	59	60
61	62	63	64	65	66	67	68	69	70
71	72	73	74	75	76	77	78	79	80
81	82	83	84	85	86	87	88	89	90
91	92	93	94	95	96	97	98	99	100

Using the idea of multiples, let's try to organize which numbers can be weighed.

Reflect

Connect

— Problem —

Number 8 _____ 2 _____.

I found the incomplete note shown above. What was written in it?

Before and after number 2, there can be words, numbers and blank spaces.

◎ Which words and numbers will complete the sentence?

Purpose | Let's try to think various ways to represent numbers.

Addition

Number 8 <u>is</u> <u>2</u> <u>added four times.</u>
(Math sentence) $8 = 2 + 2 + 2 + 2$

←——→ Same meaning

Number 8 <u>is</u> <u>2</u> <u>gathered four times.</u>
(Math sentence) $8 = 2 × 4$

Multiplication

Number 8 <u>is</u> <u>2</u> <u>subtracted from 10.</u>
(Math sentence) $8 = 10 - 2$

Subtraction

Area of a rectangle

Number 8 <u>is</u> <u>2</u> <u>multiplied by itself three times.</u>
(Math sentence) $8 = 2 × 2 × 2$

Multiplication

Number 8 <u>is</u> <u>2</u> <u>multiplied by 4.</u>

Number 8 <u>is 4 times of</u> <u>2</u> .

Can you represent number 8 using number 2?

If you turn the math sentence $8 = 2 × 4$ into words, it becomes "number 8 is 4 times 2."

Nanami

If $2 × 4$ is represented by a figure, it looks like this?

Yui

When represented in math sentences, there is no division.

Is there any way to represent it by division?

Number 8 __is 16 divided by__ 2 _____.

(Math sentence) $8 = 16 \div 2$ — Division

Number 8 __is__ 2 __larger than 6__.

(Math sentence) $8 = 2 + 6$ — Addition

Other ways :

Number 8 __is multiple of 2__ _____.

Multiple of 2: 2, 4, 6, 8 ......

Even number ⟷ Odd number

0 is also an even number.

→ Using "common multiple":
Number 8 __is a common multiple of__ 2 __and 4__.

→ Using "divisor":
Number 8 __is a divisor of__ 24.

Summary You can think various ways to represent numbers.

○ One number can be represented by the sum, difference, product, or quotient of other numbers.

○ Multiplication is performed when a number is represented as the area of a rectangle or square.

Product of the same 2 numbers — 1, 4, 9, 16......

○ If you use multiples or divisors, it will come as member of the same category inside whole numbers.

The area of a 2cm long and 4 cm wide rectangle, $2 \times 4 = 8$, is 8cm^2.

Daiki

Can it also be represented by a cube?

Hiroto

Want to connect

If 17 is represented as a product, it is only 1 and 17 as 1×17. Are there other numbers like this?

Nanami

Continue at Junior High School.

Deepen.

Number with 2 divisors.

The number 17 only has two divisors: 1 and 17. Let's find, from the numbers on the right, other whole numbers in which its only divisors are 1 and the number itself.

1	2	3	4	5	6	7	8	9	10
11	12	13	14	15	16	17	18	19	20
21	22	23	24	25	26	27	28	29	30
31	32	33	34	35	36	37	38	39	40

A number, such as 2, 3, 5, 7, 11, ..., in which its only divisors are 1 and the number itself is called a **prime number**. 1 is not a prime number.

① Use the following procedure to find the prime numbers from 1 to 100.

(1) Erase number 1.

(2) Leaving number 2, erase the multiples of 2.

(3) Leaving number 3, erase the multiples of 3.

...

In this way, out of the remaining numbers (leaving the first one) the multiples are erased.

With this method, only prime numbers remain. This is said to have been invented by the ancient Greek mathematician Eratosthenes, and is called "The Sieve of Eratosthenes."

1	2	3	4	5	6
7	8	9	10	11	12
13	14	15	16	17	18
19	20	21	22	23	24
25	26	27	28	29	30
31	32	33	34	35	36
37	38	39	40	41	42
43	44	45	46	47	48
49	50	51	52	53	54
55	56	57	58	59	60
61	62	63	64	65	66
67	68	69	70	71	72
73	74	75	76	77	78
79	80	81	82	83	84
85	86	87	88	89	90
91	92	93	94	95	96
97	98	99	100		

A whole number, such as $6 = 2 \times 3$, can be expressed as a product of prime numbers. Let's try to express 30 and 42 in the form of a product of prime numbers.

Supplementary Problems

1 Decimal Numbers and Whole Numbers

p.10~p.17

1 Let's fill in the following ☐ with numbers.

① $48.3 = 10 \times \boxed{} + 1 \times \boxed{}$

$+ 0.1 \times \boxed{}$

② $0.529 = 0.1 \times \boxed{} + 0.01 \times \boxed{}$

$+ 0.001 \times \boxed{}$

③ $9.87 = \boxed{} \times 9 + \boxed{} \times 8$

$+ \boxed{} \times 7$

④ $16.04 = \boxed{} \times 1 + \boxed{} \times 6$

$+ \boxed{} \times 4$

⑤ $1 \times 7 + 0.01 \times 5 + 0.001 \times 8$

$= \boxed{}$

⑥ $0.01 \times 3 + 0.001 \times 9 = \boxed{}$

2 Let's find 10 times and 100 times of the following numbers.

① 2.46 ② 0.507

③ 17.92 ④ 0.083

3 Let's find $\frac{1}{10}$ and $\frac{1}{100}$ of the following numbers.

① 814 ② 5.36

③ 17.09 ④ 602.5

4 Let's fill in the following ☐ with numbers.

① $\boxed{}$ times of 54.9 is 549.

② $\boxed{}$ times of 0.286 is 28.6.

③ $\frac{1}{\boxed{}}$ of 9.3 is 0.093.

④ $\frac{1}{\boxed{}}$ of 702.1 is 70.21.

5 Let's find the following numbers.

① 0.253×10

② 4.09×100

③ 0.086×100

④ $17.2 \div 10$

⑤ $9.56 \div 100$

⑥ $347.2 \div 100$

6 Let's use the five numerals 1, 3, 4, 7, 8 once and the decimal point to create the following decimal numbers.

① The number closest to 3

② The number closest to 5

7 Let's use the five numerals 0, 2, 5, 6, 9 once and the decimal point to create the following decimal numbers.

① The number closest to 1

② The number closest to 50

② Congruent Figures
p.18~p.31

❶ From the following figures, which figures are congruent?

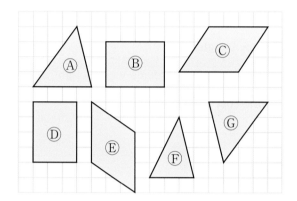

❷ The following 2 quadrilaterals are congruent. Let's answer the questions below.

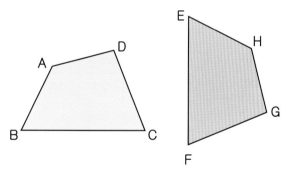

① Which is the corresponding vertex for the following vertices?

    ⓐ   vertex A   ⓑ   vertex F

② Which is the corresponding side for the following sides?

    ⓐ   side AB   ⓑ   side FG

③ Which is the corresponding angle for the following angles?

    ⓐ   angle D   ⓑ   angle E

❸ The following 2 triangles are congruent. Let's answer the questions below.

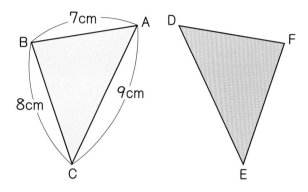

① How many cm is the length of the following sides?

    ⓐ   side DE   ⓑ   side DF

② Which angle has equal size as the following angles?

    ⓐ   angle B   ⓑ   angle D

❹ I would like to draw a triangle that is congruent to triangle ABC. Which one more thing do I need to know? Which length of a side or size of an angle?

①

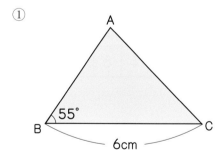

②

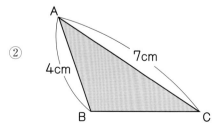

5 Let's draw a quadrilateral that is congruent to quadrilateral ABCD.

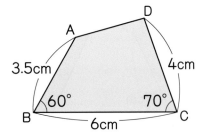

③ Proportion

p.32~p.39

1 From Ⓐ, Ⓑ, and Ⓒ, let's choose the one in which ○ is proportional to □.

Ⓐ Length of one side □ cm and area ○ cm² of a square.

One side length □(cm)	1	2	3	4	5
Area ○(cm²)	1	4	9	16	25

Ⓑ Length of one side □ cm and surrounding length ○ cm of a square.

One side length □(cm)	1	2	3	4	5
Surrounding length ○(cm)	4	8	12	16	20

Ⓒ Length □ m and weight ○ g when one meter of wire weighs 15 g.

Length □(m)	1	2	3	4	5
Weight ○(g)	15	30	45	60	75

2 Let's answer the following questions using □ number of icecream and total cost ○ yen when one icecream costs 80 yen.

① Let's represent the relationship between □ and ○ in a math sentence.

② What is the total cost when 6 icecreams were bought?

③ How many icecreams were bought if the total cost is 960 yen?

④ Mean

p.40~p.49

1 The following table shows the score for the mathematics test of 5 people in one group. Let's find the mean score.

Score in the mathematics test

Name	Sho	Tsubasa	Karin	Yuna	Miki
Score (points)	78	87	92	75	84

2 After examining the weight of 6 eggs, it became as follows.

How many g is the mean weight of one egg?

54g 59g 60g 55g 58g 53g

3 The following table represents the number of pages that Moka reads in 5 days. What is the mean number of pages that she reads in one day?

Number of pages Moka read

Day number	1st Day	2nd Day	3rd Day	4th Day	5th Day
Number of pages	26	33	18	28	30

4 The goals made by a soccer team in the last 8 games are as follows. What is the mean number of goals in one game?

2 goals   1 goal   3 goals   0 goal   1 goal
1 goal   3 goals   1 goal

⑤ Measure per Unit Quantity (1)

p.50~p.61

1 18 children are playing in sandbox A that has an area of 12 m². Also, 21 children are playing in sandbox B that has an area of 15 m². Which sandbox is more crowded?

2 From trains Ⓐ and Ⓑ, which is more crowded?

Ⓐ   A train with 6 cars and 870 passengers.

Ⓑ   A train with 9 cars and 1050 passengers.

3 The following table shows the population and area of City A, City B, City C, and City D. Let's answer the following questions.

Population and area

	Population (people)	Area (km²)
City A	43510	95
City B	57800	110
City C	39000	68
City D	57750	154

① Let's find the population density of each city. Round off the tenths place and express the answer in whole numbers.

② Which city has the highest population density?

4 Let's answer the questions about a wire that is 6 m long and weighs 390 g.

① How many grams is the weight for 1 m of wire?

② How many grams is the weight for 14 m of wire?

③ When this wire weighs 533 g, how many meters is it?

5 There are 9 pencils for 720 yen and 12 pencils for 900 yen. Can you say which pencil is more expensive? Let's compare by the price per pencil.

6 Let's answer the following questions about a car that runs 288 km using 16 L of gasoline.
① How many kilometers can this car run using 1 L of gasoline?
② How many kilometers can this car run using 22 L of gasoline?
③ How many liters of gasoline will this car use to run 468 km?

6 Multiplication of Decimal Numbers
p.64～p.78

1 Let's solve the following calculations in vertical form.
① 70×3.6
② 80×2.7
③ 5×1.7
④ 9×2.1
⑤ 17×2.4
⑥ 32×1.8

2 Considering the calculation of 90×1.6, let's fill in the following ☐ with numbers.

$$90 \times 1.6 = 90 \times 16 \div \boxed{}$$
$$= \boxed{}$$

3 There is a rectangular flowerbed with a length of 4 m and a width of 2.8 m.
How many square meters is the area of the flowerbed?

4 On the following calculations, let's place the decimal point of the product.

①
```
    2.3
 ×  3.5
  1 1 5
  6 9
  8 0 5
```

②
```
   1.6 3
 ×   1.7
 1 1 4 1
 1 6 3
 2 7 7 1
```

5 Let's solve the following calculations in vertical form.
① 3.2×2.2
② 2.9×3.8
③ 4.7×6.6
④ 7.3×5.4
⑤ 1.36×1.8
⑥ 4.13×3.7
⑦ 5.1×2.64
⑧ 6.2×4.39

6 Let's solve the following calculations in vertical form.

① 2.5 × 3.8   ② 1.6 × 4.5

③ 0.4 × 2.3   ④ 0.3 × 1.9

⑤ 3.15 × 4.6   ⑥ 5.42 × 3.5

7 The length of one side of a squared flowerbed is 4.7 m. How many square meters is the area of the flowerbed?

8 Let's solve the following calculations in vertical form.

① 2.6 × 0.4   ② 0.9 × 0.7

③ 3.82 × 0.8   ④ 0.65 × 0.2

⑤ 4.7 × 0.36   ⑥ 0.08 × 0.6

9 Let's fill in the following ☐ with inequality signs.

① 3.8 × 0.9 ☐ 3.8

② 4.12 × 1.3 ☐ 4.12

③ 0.57 × 2.4 ☐ 0.57

④ 9.8 × 0.6 ☐ 9.8

10 Let's answer about iron bars that weigh 2.8kg per meter.

① How many kilograms is the weight of an iron bar that has a length of 2.5 m?

② How many kilograms is the weight of an iron bar that has a length of 0.7 m?

11 Let's fill in the following ☐ with numbers.

① $1.3 \times 0.63 = \boxed{} \times 1.3$

② $(4.9 \times 3.5) \times 0.2$

$= 4.9 \times (3.5 \times \boxed{})$

12 Let's fill in the following ☐ with numbers.

① $6.8 \times 4.7 + 6.8 \times 5.3$

$= 6.8 \times (4.7 + \boxed{})$

$= 6.8 \times \boxed{}$

$= \boxed{}$

② $5.7 \times 2.5 - 1.7 \times 2.5$

$= (5.7 - \boxed{}) \times 2.5$

$= \boxed{} \times 2.5$

$= \boxed{}$

13 Let's improve the calculation by using the rules of operations.

① 4.7 × 4 × 2.5

② 0.6 × 7.2 × 5

③ 3.6 × 1.4 + 6.4 × 1.4

④ 9.1 × 4.6 − 4.1 × 4.6

14 Instead of multiplying a number by 1.8, a friend added 1.8 to the number and got an answer of 6.3 by mistake. What shoud have been the answer to the original problem?

7 Division of Decimal Numbers

p.79~p.95

1 Let's solve the following calculations in vertical form.

① 7 ÷ 1.4　　② 9 ÷ 1.5

③ 63 ÷ 4.5　　④ 7 ÷ 5.6

2 There is a rectangle with an area of 24 cm² and a length of 3.2 cm. How many centimeters is the width of this rectangle?

3 Let's fill in the following ☐ with numbers.

Since the quotient of 13 ÷ 2.5 is equal to the quotient of 130 ÷ ☐ , and

130 ÷ 25 = ☐ , then 13 ÷ 2.5 = ☐ .

4 Let's solve the following calculations in vertical form.

① 6.72 ÷ 2.1　　② 8.28 ÷ 1.8

③ 6.24 ÷ 2.4　　④ 9.36 ÷ 3.6

⑤ 8.4 ÷ 2.1　　⑥ 9.6 ÷ 1.6

⑦ 7.8 ÷ 1.3　　⑧ 8.1 ÷ 2.7

5 There is a 1.7 m iron bar that weighs 5.78 kg. How many kilograms is the weight of 1 m of this iron bar?

6 7.2 L of juice was poured into 1.8 L bottles. How many bottles were needed to pour in the juice?

7 Let's solve the following calculations in vertical form.

① 9.9 ÷ 2.2　　② 6.3 ÷ 3.5

③ 51.6 ÷ 2.4　　④ 55.9 ÷ 6.5

⑤ 4.42 ÷ 8.5　　⑥ 3.42 ÷ 7.6

⑦ 4.93 ÷ 5.8　　⑧ 6.08 ÷ 9.5

8 There is a rectangular flowerbed with an area of 25.3 m². When the length is 4.6 m, how many meters is the width?

9 2.6 dL of paint was used to paint a 6.5 m² wall. Let's answer the following questions.

① How many deciliters of paint were used to paint 1 m² of the wall?

② With 1 dL of paint, how many square meters of the wall can be painted?

10 Let's answer about a 1.8 m wire that weighs 22.5 g.

① How many meters is the length of this wire per gram?

② How many grams is the weight of a wire that is 1 m long?

11 Let's solve the following calculations in vertical form.

① 28 ÷ 0.4　　② 7 ÷ 0.8

③ 4.8 ÷ 0.6　　④ 3.4 ÷ 0.5

⑤ 1.2 ÷ 0.8　　⑥ 0.3 ÷ 0.2

⑦ 0.3 ÷ 0.5　　⑧ 0.1 ÷ 0.4

12 Let's fill in the following ⬚ with equality or inequality signs.

① 1.71 ÷ 0.9 ⬚ 1.71

② 2.72 ÷ 1.6 ⬚ 2.72

③ 3.5 ÷ 1.4 ⬚ 3.5

④ 0.8 ÷ 1 ⬚ 0.8

⑤ 0.58 ÷ 0.4 ⬚ 0.58

13 There is a 0.8 m pipe that weighs 1.16 kg. How many kilograms is the weight of a pipe that is 1 m long?

14 A tape that is 45 cm long is cut into pieces that are 8.5 cm long. How many 8.5 cm long pieces were cut? How many centimeters of tape remained?

15 There are 5.2 L of juice. This juice is poured into 0.6 L cups. How many cups will be filled? How many liters of juice will remain?

16 Where do you align and place the decimal point of the remainder in the division of decimal numbers in vertical form?

17 In the following division, let's find the quotient as a whole number and show the remainder.

① 22.2 ÷ 2.7   ② 28.9 ÷ 1.9

③ 7.5 ÷ 0.7   ④ 14.9 ÷ 0.6

18 Let's solve the following calculation and round off the quotient to the nearest hundredth.

① 3.1 ÷ 2.3   ② 7 ÷ 3.8

③ 89.2 ÷ 4.7   ④ 52.3 ÷ 6.8

⑤ 0.64 ÷ 3.4   ⑥ 0.32 ÷ 5.6

19 There is a car that runs 94 km with 6.5 L of gasoline. About how many kilometers does this car run with 1 L of gasoline? Let's round off the answer to the nearest tenth.

20 There is a 4.5 m long wire that weighs 35.6 g. About how many grams is the weight of a wire that is 1 m long? Let's round off the answer to the nearest tenth.

21 There is a 5.8 L oil that weighs 5.1 kg. About how many kilograms is the weight of a 1 L oil? Let's round off the answer to the nearest tenth.

22 A flowerbed is watered. Let's answer the following questions when an area of 1 m² is watered with 2.6 L of water.

① How many liters of water is needed to water an area of 8.5 m²?

② If 16.9 L of water were used, how many square meters of the flowerbed were watered?

8 Measure per Unit Quantity (2)

p.98~p.107

1 Haruma took 9 seconds to run 63 m, and Ryu took 12 seconds to run 90 m. Let's answer the following questions.

① How many meters per second is the running speed of Haruma and Ryu?

② Who ran faster?

2 Let's find the following speeds.

① There is a train that runs 300 km in 4 hours. How many kilometers per hour is the speed of this train?

② There is a car that runs 21 km in 15 minutes. How many kilometers per minute is the speed of this car?

③ There is a motorcycle that runs 420 m in 30 seconds. How many meters per second is the speed of this motorcycle?

3 Let's find the following speeds.

① How many kilometers per hour is the speed of a bus that runs at 900 m per minute? Also, how many meters per second?

② How many kilometers per minute is the speed of a train that runs at 162 km per hour? Also, how many meters per second?

③ How many meters per minute is the speed of a person that runs at 5 m per second? Also, how many kilometers per hour?

4 There is a car that runs 48 km in 40 minutes. How many kilometers per hour is the speed of this car?

5 Between a motor boat that runs at 33 km per hour and a horse that runs at 12 m per second, which one is faster?

6 Let's find the following distances.

① As for a cat that runs at 13 m per second, how many meters will it run in 6 seconds?

② As for a person that walks at 65 m per minute, how many meters will he walk in 30 minutes?

③ As for a truck that runs at 56 km per hour, how many kilometers will it travel in 3 hours?

④ As for a jet that flies at 840 km per hour, how many kilometers will it travel in 2 hours?

7 Let's find the following time.

① As for a cheetah that runs at 30 m per second, how many seconds will it take to run 195 m?

② As for a person that runs at 280 m per minute, how many minutes will it take to run 4200 m?

③ As for a bicycle that runs at 350 m per minute, how many minutes will it take to run 14 km?

④ As for a car that runs at 68 km per hour, how many hours will it take to run 204 km?

8 The distance from Miyu's house to school is 840 m. It took her a total 26 minutes to walk a round trip between her house and school. If her speed towards school was 70 m per minute, then how many meters per minute was her returning speed?

⑨ Angles of Figures

p.112~p.123

1 Let's fill in the following ☐ with numbers.

①
45°

②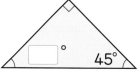
50°

Isosceles triangle

③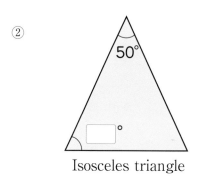
55°
60°

④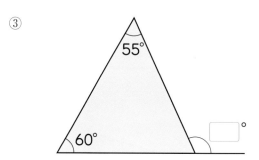
80°
140°

⑤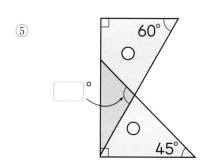
60°
45°

2 Let's fill in the ☐ with numbers.

① In any triangle, the sum of the three angles is ☐°.

② In any quadrilateral, the sum of the four angles is ☐°.

3 Let's fill in the following ☐ with numbers.

①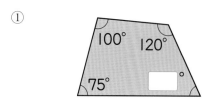
100° 120°
75°

②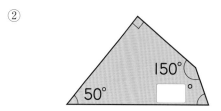
150°
50°

③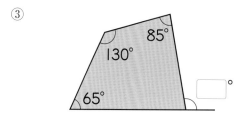
85°
130°
65°

④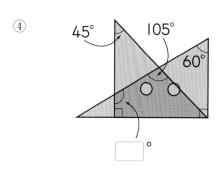
45° 105°
60°

4 In the following pentagon, let's fill in the ▢
with a number.

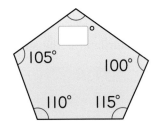

5 Let's find the sum of the six angles in a
hexagon.

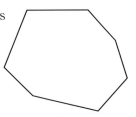

① How many diagonals
can be drawn from
one vertex?

② Considering the diagonals in ①, the
hexagon can be divided in how many
triangles?

③ How many degrees is the sum of the six
angles of a hexagon?

6 Let's find the sum of the seven angles in a
heptagon.

① How many diagonals
can be drawn from
one vertex?

② Considering the diagonals in ①, the
heptagon can be divided in how many
triangles?

③ How many degrees is the sum of the seven
angles of a heptagon?

7 Let's fill in the ▢ with numbers.

① The sum of the eight angles of a octagon is
$180° × \boxed{} = \boxed{}°$.

② The sum of the nine angles of a nonagon is
$180° × \boxed{} = \boxed{}°$.

③ The sum of all the angles of a decagon is
$180° × \boxed{} = \boxed{}°$.

8 Find the sum of the four angles of a
quadrilateral by dividing it into triangles as
shown below.

Let's fill in the ▢ with numbers.

①

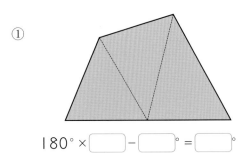

$180° × \boxed{} - \boxed{}° = \boxed{}°$

②

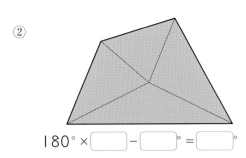

$180° × \boxed{} - \boxed{}° = \boxed{}°$

③

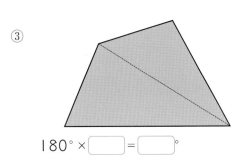

$180° × \boxed{} = \boxed{}°$

🔟 Multiples and Divisors

p.124~p.144

1 Let's categorize the following whole numbers as even or odd numbers.

15   26   98   107   253   774

2 Let's find 5 multiples of the following numbers in ascending order.

① 6   ② 15

3 Let's find 3 common multiples of the following set of numbers in ascending order.

① (2, 9)   ② (5, 10)

③ (6, 8)   ④ (4, 10)

⑤ (2, 3, 4)   ⑥ (6, 9, 12)

4 Let's find the least common multiple of the following set of numbers.

① (3, 7)   ② (8, 16)

③ (12, 15)   ④ (6, 10, 15)

5 Let's make a square by aligning 9 cm long and 15 cm wide rectangular paper in the same direction as shown on the right.

How many centimeters is the length of the side of the smallest square created?

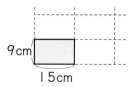

9 cm

15 cm

6 Let's find all the common divisors of the following set of numbers.

① (10, 20)   ② (12, 18)

③ (16, 40)   ④ (7, 9)

⑤ (8, 12, 16)   ⑥ (15, 30, 45)

7 Let's find the greatest common divisor of the following set of numbers.

① (14, 21)   ② (24, 36)

③ (27, 45)   ④ (12, 24, 30)

8 I would like to divide 40 red and 32 blue sheets of paper by some children such that there is no remainder and each child receives the same number. How many children will receive the sheets of paper if I divide without having remainder? Let's answer with the largest possible number of children.

Answers

❶ Decimal Numbers and Whole Numbers

❶ ① 4.8.3 ② 5.2.9 ③ 1.0.1.0.01 ④ 10.1.0.01
⑤ 7.058 ⑥ 0.039

❷ ① 10 times ···24.6   100 times ···246
② 10 times ···5.07   100 times ···50.7
③ 10 times ···179.2   100 times ···1792
④ 10 times ···0.83   100 times ···8.3

❸ ① $\frac{1}{10}$···81.4 $\frac{1}{100}$···8.14
② $\frac{1}{10}$···0.536 $\frac{1}{100}$···0.0536
③ $\frac{1}{10}$···1.709 $\frac{1}{100}$···0.1709
④ $\frac{1}{10}$···60.25 $\frac{1}{100}$···6.025

❹ ① 10 ② 100 ③ 100 ④ 10

❺ ① 2.53 ② 409 ③ 8.6 ④ 1.72 ⑤ 0.0956
⑥ 3.472

❻ ① 3.1478 ② 4.8731

❼ ① 0.9652 ② 50.269

❷ Congruent Figures

❶ Ⓐ and Ⓖ, Ⓑ and Ⓓ, Ⓒ and Ⓔ

❷ ① ⓐ vertex H ⓑ vertex C ② ⓐ side HE ⓑ side CD
③ ⓐ angle G ⓑ angle B

❸ ① ⓐ 9 cm ⓑ 7 cm ② ⓐ angle F ⓑ angle A

❹ ① length of side AB or size of angle C
② length of side BC or size of angle A

❺ (omitted)

❸ Proportion

❶ Ⓑ, Ⓒ

❷ ① 80 × □ = ○ ② 480 yen ③ 12 icecreams

❹ Mean

❶ (78 + 87 + 92 + 75 + 84) ÷ 5 = 83.2   83.2 points

❷ (54 + 59 + 60 + 55 + 58 + 53) ÷ 6 = 56.5   56.5 g

❸ (26 + 33 + 18 + 28 + 30) ÷ 5 = 27   27 pages

❹ (2 + 1 + 3 + 0 + 1 + 1 + 3 + 1) ÷ 8 = 1.5   1.5 goals

❺ Measure per Unit of Quantity (1)

❶ Sandbox A   18 ÷ 12 = 1.5
Sandbox B   21 ÷ 15 = 1.4   Sandbox A

❷ Ⓐ 870 ÷ 6 = 145
Ⓑ 1050 ÷ 9 = 116.6···   Train Ⓐ

❸ ① City A 43510 ÷ 95 = 458   458 people
City B 57800 ÷ 110 = 525.4···   525 people
City C 39000 ÷ 68 = 573.5···   574 people
City D 57750 ÷ 154 = 375   375 people
② City C

❹ ① 390 ÷ 6 = 65   65 g ② 65 × 14 = 910   910 g
③ 533 ÷ 65 = 8.2   8.2 m

❺ 720 ÷ 9 = 80
900 ÷ 12 = 75   9 pencils for 720 yen

❻ ① 288 ÷ 16 = 18   18 km
② 18 × 22 = 396   396 km
③ 468 ÷ 18 = 26   26 L

❻ Multiplication of Decimal Numbers

❶ ① 252 ② 216 ③ 8.5 ④ 18.9 ⑤ 40.8
⑥ 57.6

❷ 10.144

❸ 4 × 2.8 = 11.2   11.2 m²

❹
①
```
   2.3
 × 3 5
 1 1 5
 6 9
 8.0 5
```
②
```
   1.6 3
 ×   1 7
 1 1 4 1
 1 6 3
 2.7 7 1
```

❺ ① 7.04 ② 11.02 ③ 31.02 ④ 39.42
⑤ 2.448 ⑥ 15.281 ⑦ 13.464 ⑧ 27.218

❻ ① 9.5 ② 7.2 ③ 0.92 ④ 0.57 ⑤ 14.49
⑥ 18.97

❼ 4.7 × 4.7 = 22.09   22.09 m²

❽ ① 1.04 ② 0.63 ③ 3.056
④ 0.13 ⑤ 1.692 ⑥ 0.048

❾ ① < ② > ③ > ④ <

❿ ① 2.8 × 2.5 = 7   7 kg
② 2.8 × 0.7 = 1.96   1.96 kg

⓫ ① 0.63 ② 0.2

⓬ ① 5.3.10.68 ② 1.7.4.10

⓭ ① 47 ② 21.6 ③ 14 ④ 23

⓮ 6.3 − 1.8 = 4.5   4.5 × 1.8 = 8.1   8.1

❼ Division of Decimal Numbers

❶ ① 5 ② 6 ③ 14 ④ 1.25

❷ 24 ÷ 3.2 = 7.5   7.5 cm

❸ 25.5.2.5.2

❹ ① 3.2 ② 4.6 ③ 2.6 ④ 2.6 ⑤ 4
⑥ 6 ⑦ 6 ⑧ 3

❺ 5.78 ÷ 1.7 = 3.4   3.4 kg

6 7.2 ÷ 1.8 = 4   4 bottles

7 ① 4.5 ② 1.8 ③ 21.5 ④ 8.6
   ⑤ 0.52 ⑥ 0.45 ⑦ 0.85 ⑧ 0.64

8 25.3 ÷ 4.6 = 5.5   5.5 m

9 ① 2.6 ÷ 6.5 = 0.4   0.4 dL
   ② 6.5 ÷ 2.6 = 2.5   2.5 m²

10 ① 1.8 ÷ 22.5 = 0.08   0.08 m
   ② 22.5 ÷ 1.8 = 12.5   12.5 g

11 ① 70 ② 8.75 ③ 8 ④ 6.8
   ⑤ 1.5 ⑥ 1.5 ⑦ 0.6 ⑧ 0.25

12 ① > ② < ③ < ④ = ⑤ >

13 1.16 ÷ 0.8 = 1.45   1.45 kg

14 45 ÷ 8.5 = 5 remainder 2.5   5 pieces, 2.5 cm remained

15 5.2 ÷ 0.6 = 8 remainder 0.4   8 cups, 0.4 L will remain

16 In the decimal point of the dividend

17 ① 8 remainder 0.6 ② 15 remainder 0.4
   ③ 10 remainder 0.5 ④ 24 remainder 0.5

18 ① 1.35 ② 1.84 ③ 18.98 ④ 7.69 ⑤ 0.19
   ⑥ 0.06

19 94 ÷ 6.5 = 14.46⋯   about 14.5 km

20 35.6 ÷ 4.5 = 7.91⋯   about 7.9 g

21 5.1 ÷ 5.8 = 0.87⋯   about 0.9 kg

22 ① 2.6 × 8.5 = 22.1   22.1 L
   ② 16.9 ÷ 2.6 = 6.5   6.5 m²

8 Measure per Unit Quantity (2)

1 ① Haruma 63 ÷ 9 = 7   7 meters per second
      Ryu 90 ÷ 12 = 7.5   7.5 meters per second
   ② Ryu

2 ① 300 ÷ 4 = 75   75 kilometers per hour
   ② 21 ÷ 15 = 1.4   1.4 kilometers per minute
   ③ 420 ÷ 30 = 14   14 meters per second

3 ① 54 kilometers per hour, 15 meters per second
   ② 2.7 kilometers per minute, 45 meters per second
   ③ 300 meters per minute, 18 kilometers per hour

4 48 ÷ 40 × 60 = 72   72 kilometers per hour

5 Motor boat 33000 ÷ 60 = 550   550 meters per minute
   Horse 12 × 60 = 720   720 meters per minute   Horse

6 ① 13 × 6 = 78   78 m
   ② 65 × 30 = 1950   1950 m
   ③ 56 × 3 = 168   168 km
   ④ 840 × 2 = 1680   1680 km

7 ① 195 ÷ 30 = 6.5   6.5 seconds
   ② 4200 ÷ 280 = 15   15 minutes
   ③ 14000 ÷ 350 = 40   40 minutes
   ④ 204 ÷ 68 = 3   3 hours

8 840 ÷ 70 = 12
   840 ÷ (26 − 12) = 60   60 meters per minute

9 Angles of Figures

1 ① 45 ② 65 ③ 115 ④ 60 ⑤ 105

2 ① 180 ② 360

3 ① 65 ② 70 ③ 100 ④ 120

4 110

5 ① 3 diagonals ② 4 ③ 720°

6 ① 4 diagonals ② 5 ③ 900°

7 ① 6. 1080 ② 7. 1260 ③ 8. 1440

8 ① 3. 180. 360 ② 4. 360. 360 ③ 2. 360

10 Multiples and Divisors

1 Even numbers⋯26. 98. 774   Odd numbers⋯15. 107. 253

2 ① 6. 12. 18. 24. 30 ② 15. 30. 45. 60. 75

3 ① 18. 36. 54 ② 10. 20. 30 ③ 24. 48. 72
   ④ 20. 40. 60 ⑤ 12. 24. 36 ⑥ 36. 72. 108

4 ① 21 ② 16 ③ 60 ④ 30

5 45 cm

6 ① 1. 2. 5. 10 ② 1. 2. 3. 6 ③ 1. 2. 4. 8
   ④ 1 ⑤ 1. 2. 4 ⑥ 1. 3. 5. 15

7 ① 7 ② 12 ③ 9 ④ 6

8 8 children

Words and symbols from this book.

Angles of Figures

▼ will be used in pages 117 and 118.

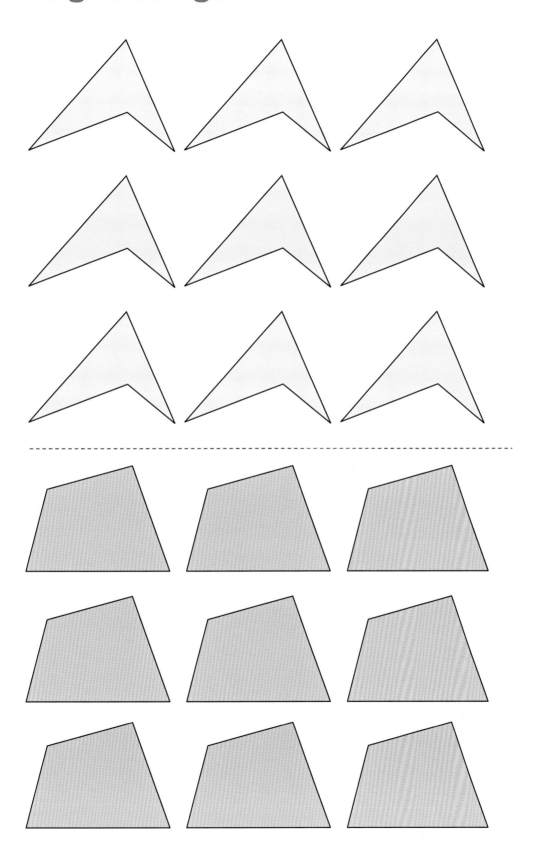

Congruent Figures

▼ will be used in page 19.

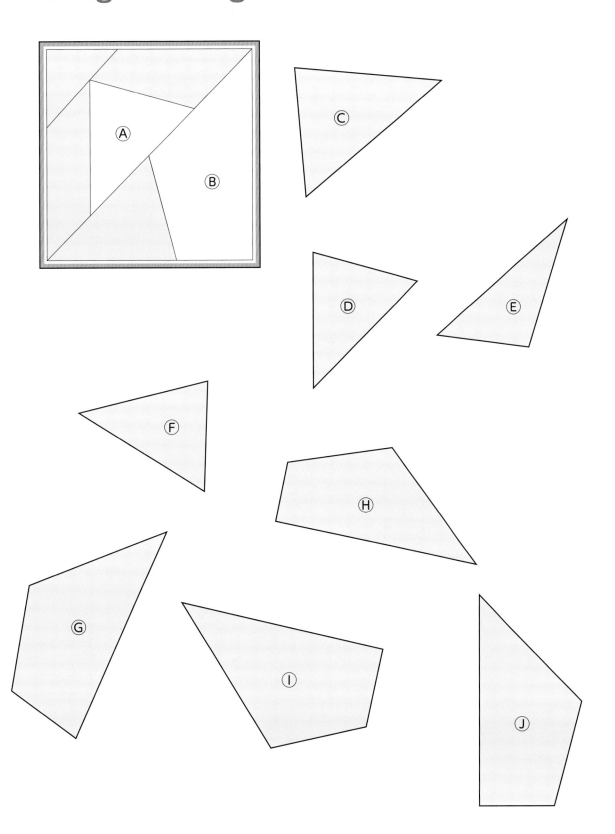

Memo

Memo

Memo

Editorial for English Edition:

Study with Your Friends, Mathematics for Elementary School

5th Grade, Vol.1, Gakko Tosho Co.,Ltd., Tokyo, Japan [2020]

Chief Editor:

Masami Isoda (University of Tsukuba, Japan), Aki Murata (University of Florida, USA)

Advisory Board:

Abrham Arcavi (Weizmann Institute of Science, Israel), Aida Istino Yap (University of the Philippines, Philippines), Alf Coles (University of Bristol, UK), Bundit Thipakorn (King Mongkut's University of Technology Thonburi, Thailand), Fou-Lai Lin (National Taiwan Normal University, Taiwan), Hee-Chan Lew (Korean National University of Education, Korea), Lambas (Ministry of Education and Culture, Indonesia), Luc Trouche (Ecole Normale Supérieure de Lyon, France), Maitree Inprasitha (Khon Kaen University, Thailand), Marcela Santillán (Universidad Pedagógica Nacional, Mexico), María Jesús Honorato Errázuriz (Ministry of Education, Chile), Raimundo Olfos Ayarza (Pontificia Universidad Católica de Valparaíso, Chile), Rogin Huang (Middle Tennessee State University, USA), Suhaidah Binti Tahir (SEAMEO RECSAM, Malaysia), Sumardyono (SEAMEO QITEP in Mathematics, Indonesia), Toh Tin Lam (National Institute of Education, Singapore), Toshikazu Ikeda (Yokohama National University, Japan), Wahyudi (SEAMEO Secretariat, Thailand), Yuriko Yamamoto Baldin (Universidade Federal de São Carlos, Brazil)

Editorial Board:

Abolfazl Rafiepour (Shahid Bahonar University of Kerman, Iran), Akio Matsuzaki (Saitama University, Japan), Cristina Esteley (Universidad Nacional de Córdoba, Argentina), Guillermo P. Bautista Jr. (University of the Philippines, Philippines), Ivan Vysotsky (Moscow Center for Teaching Excellence, Russia), Kim-Hong Teh (SEAMEO RECSAM, Malaysia), María Soledad Estrella (Pontificia Universidad Católica de Valparaiso, Chile), Narumon Changsri (Khon Kaen University, Thailand), Nguyen Chi Thanh (Vietnam National University, Vietnam), Onofre Jr Gregorio Inocencio (Foundation to Upgrade the Standard of Education, Philippines), Ornella Robutti (Università degli Studi di Torino, Iraly), Raewyn Eden (Massey University, New Zealand), Roberto Araya (Universidad de Chile, Chile), Soledad Asuncion Ulep (University of the Philippines, Philippines), Steven Tandale (Department of Education, Papua New Guinea), Tadayuki Kishimoto (Toyama University, Japan), Takeshi Miyakawa (Waseda University, Japan), Tenoch Cedillo (Universidad Pedagógica Nacional, Mexico), Ui Hock Cheah (IPG Pulau Pinang, Malaysia), Uldarico Victor Malaspina Jurado (Pontificia Universidad Católica del Perú, Peru), Wahid Yunianto (SEAMEO QITEP in Mathematics, Indonesia), Wanty Widjaja (Deakin University, Australia)

Translators:

Masami Isoda, Solis Worsfold Diego, Tsuyoshi Nomura (University of Tsukuba, Japan)

Hideo Watanabe (Musashino University, Japan)